MONTGOMERY COLLEGE LIBRARY
GERMANTOWN CAMPUS

Statistics in the Real World

Statistics in the Real World: A Book of Examples

Richard J. Larsen
and
Donna Fox Stroup

Macmillan Publishing Co., Inc.
New York
Collier Macmillan Publishers
London

Copyright © 1976, Macmillan Publishing Co., Inc.
Printed in the United States of America

All rights reserved. No part of this book may be reproduced or transmitted in any form or by any means, electronic or mechanical, including photocopying, recording, or any information storage and retrieval system, without permission in writing from the Publisher.

Macmillan Publishing Co., Inc.
866 Third Avenue, New York, New York 10022

Collier Macmillan Canada, Ltd.

Printing: 2 3 4 5 6 7 8   Year: 6 7 8 9 0 1 2

Foreword

"You haven't told me yet," said Lady Nuttal, "what it is your fiancé does for a living."

"He's a statistician," replied Lamia, with an annoying sense of being on the defensive.

Lady Nuttal was obviously taken aback. It had not occurred to her that statisticians entered into normal social relationships. The species, she would have surmised, was perpetuated in some collateral manner, like mules.

"But Aunt Sara, it's a very interesting profession," said Lamia warmly.

"I don't doubt it," said her aunt, who obviously doubted it very much. "To express anything important in mere figures is so plainly impossible that there must be endless scope for well-paid advice on how to do it. But don't you think that life with a statistician would be rather, shall we say, humdrum?"

Lamia was silent. She felt reluctant to discuss the surprising depth of emotional possibility which she had discovered below Edward's numerical veneer.

"It's not the figures themselves," she said finally, "it's what you do with them that matters."

K. A. C. Manderville
The Undoing of Lamia Gurdleneck

Preface

Until fairly recently, statistics was not a subject that attracted much attention on college campuses. The number of students taking statistics courses was relatively small, and the selection of courses quite limited. Very often, if a college had any statistics courses at all, they were highly theoretical and demanded the sort of background that only a junior or senior math major might have. The typical liberal arts major was left out.

Today, all of this has changed. As science became more quantitative, the need for statistics—at all levels—became more obvious. New courses were developed, old ones were revised, and everywhere enrollments mushroomed. At many schools, departments of psychology, education, biology, economics, and business list one semester of statistics as a requirement. And many students now have the option of taking probability and statistics rather than college algebra or calculus as a means of fulfilling a math distribution requirement.

The vast majority of these new courses aimed at the liberal arts major are taught at the precalculus level. Of necessity, such courses must emphasize general principles and applications as opposed to rigorous theory. For teachers not having any extensive, first-hand experience in statistics, this can pose a serious problem. Explaining the *principles* of statistics is fairly easy, but showing how those principles are actually applied can be quite difficult. Dice games, card problems, and made-up examples about test scores or weight gains are much too unrealistic and trivial in scope to do this effectively. It is the authors' firm conviction that the only way to convey the relevance of statistics is through the use of *real* data taken from *real* experiments.

There are fifty examples in this book. All are based on recently published research in anthropology, psychology, political science, geology, medicine, sociology, economics, biology, and even such nonstatistical areas as literature, history, and law. The examples have been divided into six chapters that parallel the usual sequence of topics presented in one- and two-semester introductory courses:

Chapter 1 Characterizing Variability
Chapter 2 The One-sample Problem
Chapter 3 The Two-sample Problem

Chapter 4 The Paired-data Problem
Chapter 5 The Correlation Problem
Chapter 6 Nonparametric Methods

As much as possible, the chapters are independent of one another, as are the examples within a chapter. Consequently, the material need not be covered in its entirety nor in the same order in which it is presented here.

All the examples have the same basic format. Each begins with a one- or two-paragraph nonstatistical introduction followed by a one-sentence statistical objective. After that comes a procedure section describing how the observations were collected, on whom, and whether any special difficulties were encountered. Sometimes a full analysis accompanies the data but more typically only the beginning is presented. Questions throughout each example help the student complete the analysis with a minimum of formal direction. Space is provided in the book for the student to record his answers.

We believe this material has four basic objectives:

1. To show how statistical principles are put into practice
2. To stimulate class discussions
3. To encourage students to analyze experimental procedures—and conclusions—critically
4. To provide a pool of extra problems for homework, class lectures, or exams

Above all, though, we hope these examples will make statistics easier to teach and more enjoyable to learn. If they can do that, the book will have been a success.

We would like to express our gratitude to the reviewers for their helpful comments, to J. Edward Neve for his editorial assistance, and to all the researchers whose data we have used.

R. J. L.
D. F. S.

Contents

List of Symbols   xiii

List of Definitions   xv

Chapter 1   Characterizing Variability   1

    1.1   Introduction   1
    1.2   Comparing two or more sets of data   4
        (Examples 1.1–1.3)

          Example 1.1   "Will the *real* Quintus Curtius Snodgrass..."   4
          Example 1.2   Politics and academia   10
          Example 1.3   Etruscan skulls   13

    1.3   Characterizing a single set of data   19
        (Examples 1.4–1.6)

          Example 1.4   Fingerprints   21
          Example 1.5   Making decisions   26
          Example 1.6   Bird calls   32

    1.4   Interpreting the unexpected   36
        (Example 1.7)

          Example 1.7   Comparing headache remedies   36

Chapter 2   The One-sample Problem   41

    2.1   Introduction   41

Contents

2.2 Continuous data: the Student t distribution   42
(Examples 2.1–2.7)

Example 2.1	Experimental esthetics	44
Example 2.2	Did she or didn't she?	52
Example 2.3	Polygraphs	57
Example 2.4	The Transylvania effect	59
Example 2.5	Geochronometry	65
Example 2.6	Echolocation	70
Example 2.7	Scientific discoveries	76

2.3 Binomial data: the normal approximation   79
(Examples 2.8–2.12)

Example 2.8	Handwriting analysis	81
Example 2.9	Death months	88
Example 2.10	Honeybees and paper flowers	94
Example 2.11	A Gallup poll	97
Example 2.12	The spiral aftereffect	101

Chapter 3   The Two-sample Problem   105

3.1 Introduction   105
3.2 The two-sample t test   106
(Examples 3.1–3.5)

Example 3.1	Statistics and the coin collector	107
Example 3.2	Quintus Curtius Snodgrass revisited	113
Example 3.3	Hospital carpeting	120
Example 3.4	The Thematic Apperception Test	124
Example 3.5	Walking exercises for the newborn	128
Example 3.6	Glacial flow	132

3.3 A two-sample test for proportions   137
(Examples 3.7–3.8)

Example 3.7	Bumper stickers	138
Example 3.8	Nightmare sufferers	143

Chapter 4   The Paired-data Problem   147

4.1 Introduction   147

4.2	The paired t test (Examples 4.1–4.7)		149
	Example 4.1	"In lane number one…"	151
	Example 4.2	ESP and hypnosis	156
	Example 4.3	Glaucoma and cornea thickness	160
	Example 4.4	Bee stings	163
	Example 4.5	Drug therapy for learning problems	166
	Example 4.6	Building a better mousetrap	169
	Example 4.7	How to stop smoking	173

Chapter 5 The Correlation Problem — 177

5.1	Introduction		177
5.2	The least squares line and the sample correlation coefficient (Examples 5.1–5.4)		178
	Example 5.1	Radioactive contamination	181
	Example 5.2	Calibrating a cricket	187
	Example 5.3	Future shock	190
	Example 5.4	Do the feathers make the bird?	194
5.3	The χ^2 test (Examples 5.5–5.9)		197
	Example 5.5	Agonistic behavior in mice	198
	Example 5.6	The sheep and goat effect	202
	Example 5.7	A famous medical experiment	205
	Example 5.8	Delinquency and birth order	207
	Example 5.9	The psychology of small cars	211

Chapter 6 Nonparametric Methods — 215

6.1	Introduction		215
6.2	Rank tests		217
	Example 6.1	The politics of war	218
	Example 6.2	Preening behavior in *Drosophila Melanogaster*	223
	Example 6.3	U.S. labor disputes	226
	Example 6.4	Suicide rates	230
	Example 6.5	Adolescent problems	235
	Example 6.6	Factors affecting well productivity	238
	Example 6.7	The 1969 draft lottery	241

List of Symbols

a The y-intercept in a least squares line, $y = a + bx$

A Alternative hypothesis

b The slope in a least squares line, $y = a + bx$

C.V.   Coefficient of variation

exp   Expected frequency (chi square problems)

H Null hypothesis

$\mu\,(\mu_X, \mu_Y, \mu_d)$ Population mean (continuous data)

μ_0 A particular value of the population mean

$\tilde{\mu}$ Population median

$n\,(m, n_X, n_Y)$ Sample size

obs   Observed frequency (chi square problems)

$p\,(p_X, p_Y)$ Probability of success in a binomial trial

p_0 A particular value for p

P Significance level ($P = .05$, $P = .01$, and so forth)

$P(E)$ The probability that the event E occurs

$P_X\,(P_Y)$ Population distribution

ρ True correlation coefficient

r Sample correlation coefficient

r_s Spearman rank correlation coefficient

$s\,(s_X, s_Y, s_d)$ Sample standard deviation

s_g Grouped standard deviation

s_p Pooled standard deviation

$\sigma\,(\sigma_X, \sigma_Y)$ Population standard deviation

$\bar{x}\,(\bar{y}, \bar{d})$ Sample mean

$\bar{x}_g$ Grouped mean

$\bar{x}_w$ Weighted mean

$\overline{X}^*\,((\overline{X} - \overline{Y})^*)$ Critical value for hypothesis test involving μ (μ_X and μ_Y)

$(X/n)^*$ Critical value for hypothesis test involving p

Z Standard normal random variable

List of Definitions

alternative hypothesis

One of the two conflicting statements about the value of a parameter that make up a hypothesis test. The other is the *null hypothesis*. The alternative hypothesis always represents a range of possible parameter values. It may be either *one sided* ($A: \mu < 15$) or *two sided* ($A: \mu \neq 15$).

bell shape

A frequently encountered data pattern. Values toward the center of the response scale are the ones that occur the most often. Deviations in either direction away from the center are less common. The distribution is symmetric.

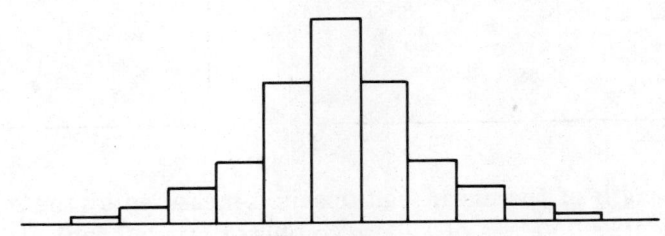

Many of the standard inference procedures (for example, the *t test*) require that the data come from a population whose distribution is more or less bell-shaped.

Bernoulli trial

An experiment whose outcome is recorded as either a "success" or a "failure." The probability of a success is denoted by the parameter p. Many real-life experiments can be thought of as consisting of a series of Bernoulli trials. The measured response in these experiments is X, the total number of successes. If the trials are independent of one another, and if the value for p remains constant, the probabilistic behavior of X is described by the *binomial distribution*.

Central Limit Theorem

This theorem states that the behavior of sample means is approximated by a normal curve (subject only to the condition that the observations are a random sample from a population distribution whose variance is finite). Without the Central Limit Theorem the principles of inference would be very difficult to put into practice. In many ways, it is the single most important result in all of statistics.

confidence interval

A range of values having a high a priori probability (usually .90 or .95) of "containing" an unknown parameter. Confidence intervals are often used in inference situations where a status quo value for the parameter in question is unavailable. If such a value were available, the analysis would call for a *hypothesis test* rather than a confidence interval.

correlation

The tendency for two jointly distributed variables to follow a linear relationship. If large values of X tend to be associated with large values of Y the correlation is said to be *positive*. If large values of X tend to be associated with small values of Y the correlation is said to be *negative*.

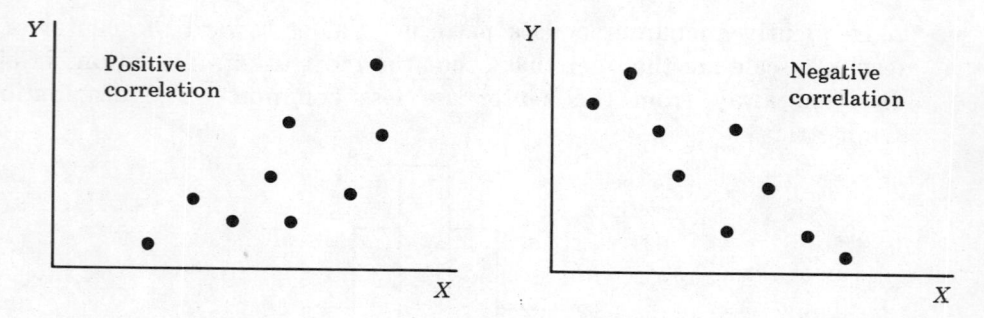

The strength of the linear relationship is measured by the *sample correlation coefficient*, r. Values of r close to plus or minus 1 indicate a strong positive or negative correlation, respectively. Values of r close to 0 indicate a weak correlation. For any set of points, $-1 \leqslant r \leqslant +1$.

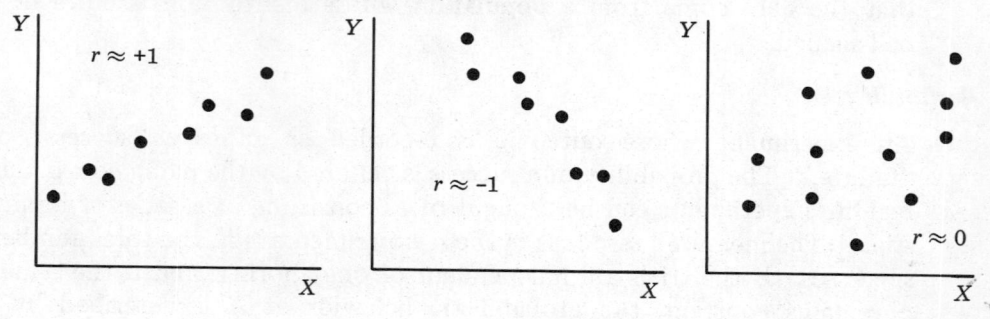

critical value

The point along the test statistic axis that separates the *acceptance region* from the *rejection region*. Only if the test statistic is more extreme than the critical value will the null hypothesis be rejected. The placement of the critical value is determined by the nature of the alternative hypothesis and the magnitude of the level of significance.

cumulative frequency polygon

A graph on which cumulative frequencies (on the vertical axis) are plotted against upper class boundaries (on the horizontal axis). Consecutive points are connected with straight lines.

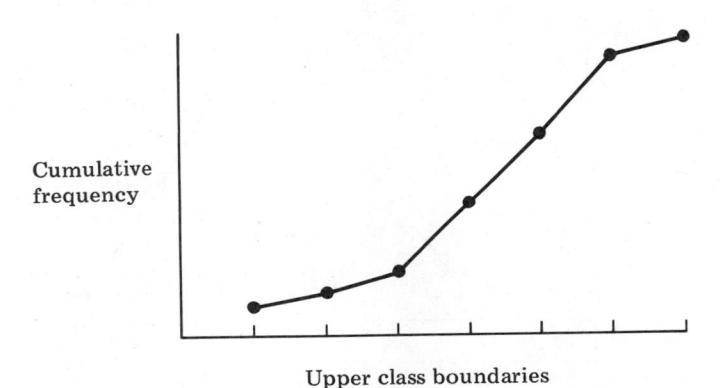

degrees of freedom

A parameter of the t, F, and χ^2 distributions. For applications of the t and F distributions, the number of degrees of freedom is related to sample size. In applications of the χ^2 distribution the number of degrees depends on the number of rows and columns in a contingency table. But whatever the situation, degrees of freedom are used to indicate which particular t or F or χ^2 distribution best describes the behavior of the test statistic.

dispersion

One of the basic characteristics of any set of numerical data. Dispersion refers to the degree of scatter among the observations—that is, the extent to which they are not all the same. It is usually measured in terms of the *standard deviation*.

double blind

A way of conducting an experiment so that neither the subject nor the person measuring the response knows which treatment the subject has been given. The purpose of doing a study double blind is to prevent the introduction of biases in the response variable, either by the subject or by the experimenter.

frequency distribution

A tabular listing showing the number of sample observations belonging to each one of a specified set of classes. The construction of a frequency distribution is often the first step taken in summarizing a set of raw data. If a more graphical format is desired, the same information can be presented as a *histogram* or *frequency polygon*.

frequency polygon

A graph on which the frequency of a class if represented by a point plotted above the midpoint of that class. Consecutive points are connected with straight lines.

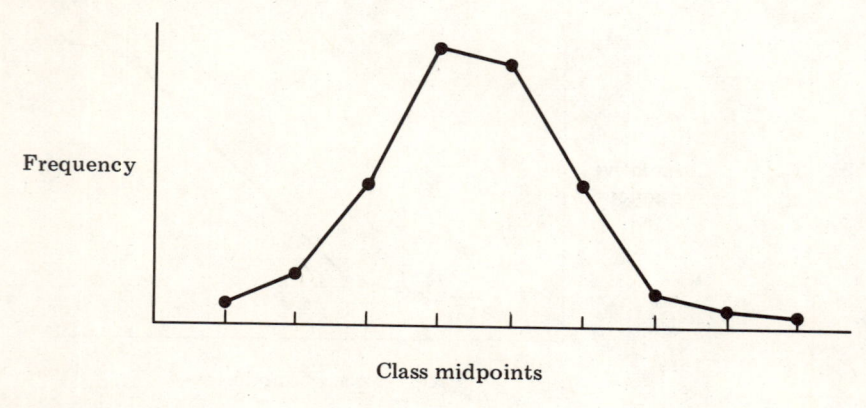

grouped mean

A mean computed from a set of data that has already been grouped into classes. Each observation in a given class is assumed to have a numerical value equal to the midpoint of that class. A *grouped standard deviation* can be computed in the same way.

grouped standard deviation

(See *grouped mean*.)

histogram

A graph in which the frequency of a class is represented by the height of a bar.

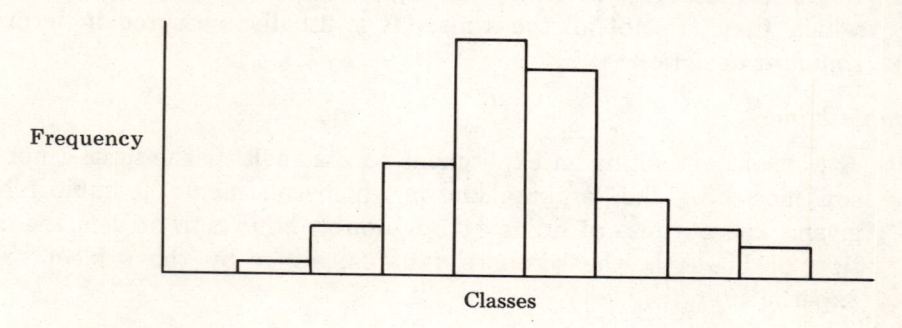

interquartile range

    A measure of dispersion defined to be the difference between the 75th and 25th *percentiles*. A large value for the interquartile range indicates that the data are highly variable.

least squares criterion

    A widely used criterion for deciding which straight line best fits a given set of data. According to the criterion, the best-fitting line, $y = a + bx$, through n points, $(x_1, y_1), (x_2, y_2), \ldots, (x_n, y_n)$, is the one that minimizes the sum of the squared vertical deviations of the points from the line—that is,

$$\sum_{i=1}^{n} (y_i - a - bx_i)^2.$$

    Formulas for the minimizing values of a and b are given in the *Least Squares Theorem*. The line satisfying the least squares criterion is called the *least squares regression line*.

level of significance

    The probability of committing a *Type I error*. In hypothesis-testing situations the null hypothesis is accepted unless the data show it to be false beyond all reasonable doubt. The level of significance provides a numerical definition of what constitutes reasonable doubt. In most instances it is set by the experimenter at either .01 or .05.

linear relationship

    A relationship between two variables that can be adequately approximated by a straight line.

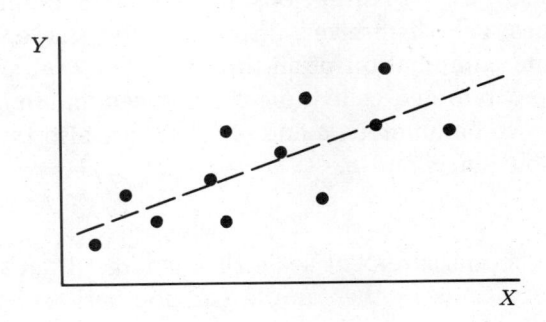

    All other functional relationships are referred to as *nonlinear* (or *curvilinear*).

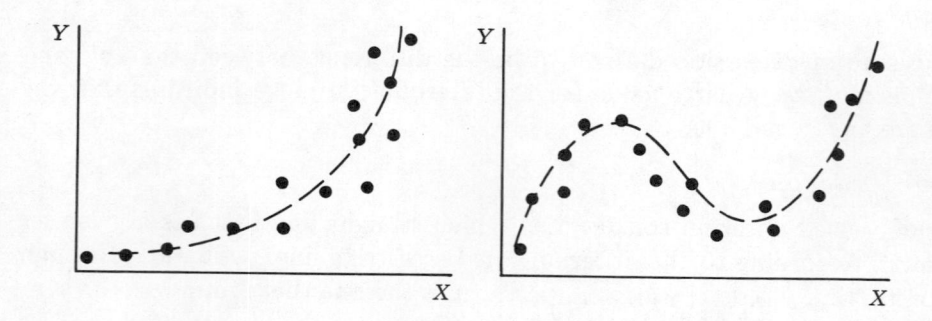

location

Like dispersion, one of the basic characteristics of statistical variation. Location refers to central tendency—that is, the position along the measurement scale where the observations tend to lie. It is usually expressed in terms of the *sample mean*.

mathematical model

A probability function that is used to approximate the behavior of a given phenomenon. The single most important mathematical model is the *normal curve*.

null hypothesis

For one-sample problems, a statement that the parameter in question equals a specified value (H: $\mu = \mu_0$). If two or more samples (and parameters) are involved, the null hypothesis frequently states that the parameters are equal (H: $p_X = p_Y$). Whatever the situation, the null hypothesis reflects the status quo (see *alternative hypothesis*).

parameter

A constant appearing in the probability function that defines a given family of distributions. Two distributions in the same family having different parameter values will, themselves, differ either in shape, location, or dispersion (or in some combination of all three). In the t, F, and χ^2 distributions, there is a single parameter, called *degrees of freedom*. In the normal distribution, there are two parameters, μ and σ. There are also two parameters in the binomial distribution, n and p.

percentile

A value along the measurement scale that is larger than a certain percentage of all the observations in the sample (or population). In general, the pth percentile, x_p, is the number such that $P(X < x_p) = p/100$. The 25th, 50th, and 75th percentiles are sometimes referred to as *quartiles*. The 50th percentile is also known as the *median*.

pooled standard deviation

An estimate of dispersion based on the information contained in two or more independent random samples. In a two-sample problem, the assumption is generally made that $\sigma_X = \sigma_Y (= \sigma)$. The pooled standard deviation is an estimate of σ. Numerically, it equals the square root of a weighted average of the two sample variances.

population distribution

The set of all possible measurements of a certain kind. The population distribution is what the sample data are presumably representing. Its mean and standard deviation are denoted by μ and σ, respectively.

random variable

A function, X, that associates a number with each possible outcome of an experiment. If an experiment were replicated many times, the outcomes would vary and the values of the random variable would vary. The resulting distribution is known as the *probability distribution of the random variable X*.

regression line

(See *least squares criterion*.)

sample distribution

The set of measurements made on the sample subjects. The mean and standard deviation of the sample distribution are denoted by $\bar{x}$ and s, respectively. The objective of every inference procedure is to generalize from what was true in the sample distribution to what might be true in the population distribution.

statistic

As opposed to a parameter, any quantity computed solely from the sample data. In hypothesis testing, the particular form of a statistic as it appears in the decision rule is known as the *test statistic*.

Type I error

Rejecting the null hypothesis when the null hypothesis is actually true. The probability of committing a Type I error is the level of significance, P.

Type II error

Accepting the null hypothesis when the alternative hypothesis is true. The probability of committing a Type II error is denoted by β. For every value of the parameter included under the alternative hypothesis, there is a different value for β. A graph of β versus the presumed value for the parameter is known as an *operating characteristic curve*.

A pinch of probably is worth a pound of perhaps.

Thurber

Chapter 1

Characterizing Variability

1.1 Introduction

To an ever-increasing extent, statistics is becoming the workhorse of the sciences. It benefits the astronomer as well as the zoologist, the economist as well as the psychologist. It can be theoretical or applied; indeed, although some statisticians devote all their efforts to proving theorems, others specialize in analyzing data. Some do both. But for all its diversity, what statistics is all about can be summed up in a single word—*variation*. Beneath the veneer of biology or sociology or mathematics, every statistical analysis is nothing but an attempt to cope with the fact that repeated measurements vary, even when taken under the most carefully controlled conditions. Where that variability comes from, whether it can be reduced, and how it can be interpreted are the questions a statistician tries to answer.

In the examples in this chapter we take a first look at the nature of variation. And what is perhaps just as important, we will begin to see how statistics can be applied in even the most unlikely of settings. We will use statistics, for example, to help unravel the mystery surrounding Quintus Curtius Snodgrass, a man whose true identity remains to this day a historical and literary puzzle; we will see what kinds of factors influence people when they make decisions; we will learn how *not* to test whether Bufferin is better than "just plain aspirin," and we will even examine a theory about the origin of the Etruscans. First, though, it may be helpful to review some basic concepts.

2 Statistics in the Real World

In characterizing the way a set of measurements $(x_1, x_2, \ldots, x_n)$ vary, we generally focus on three things:

1. The overall *shape* of the response distribution.

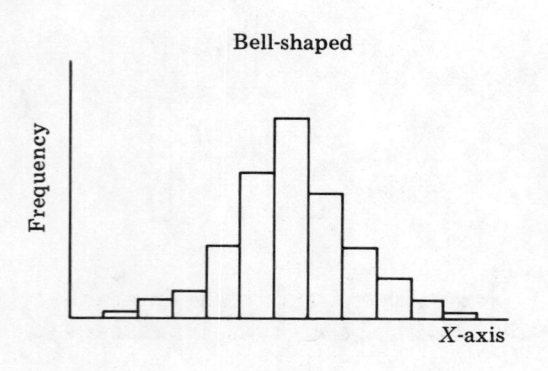

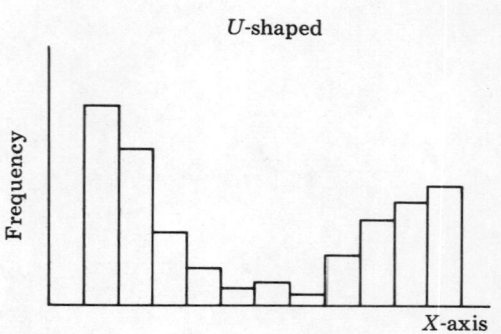

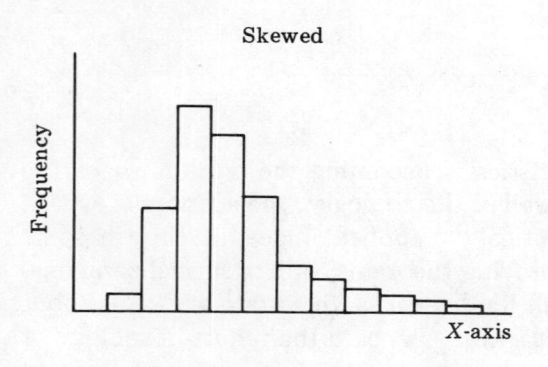

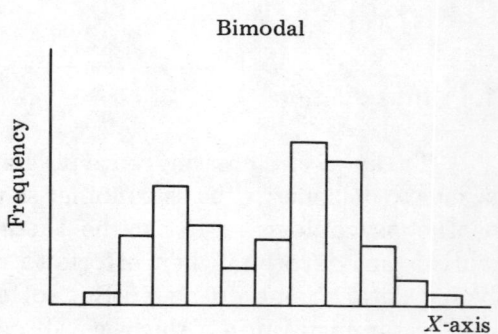

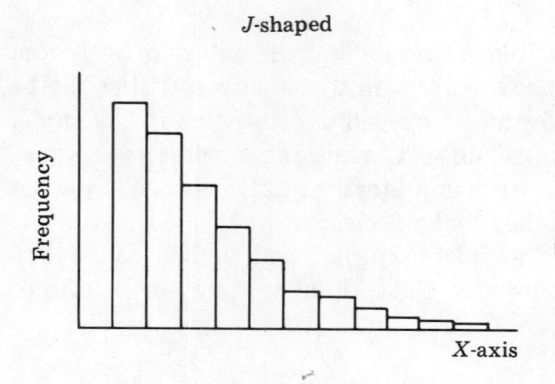

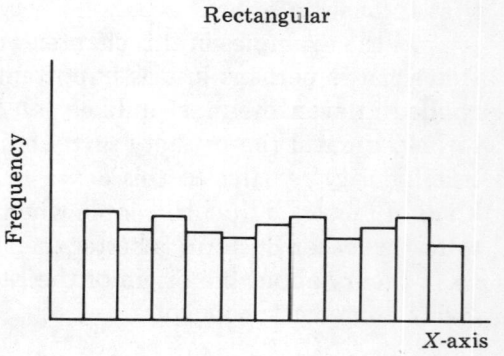

2. Where the observations tend to lie along the horizontal axis—that is, their *location*.

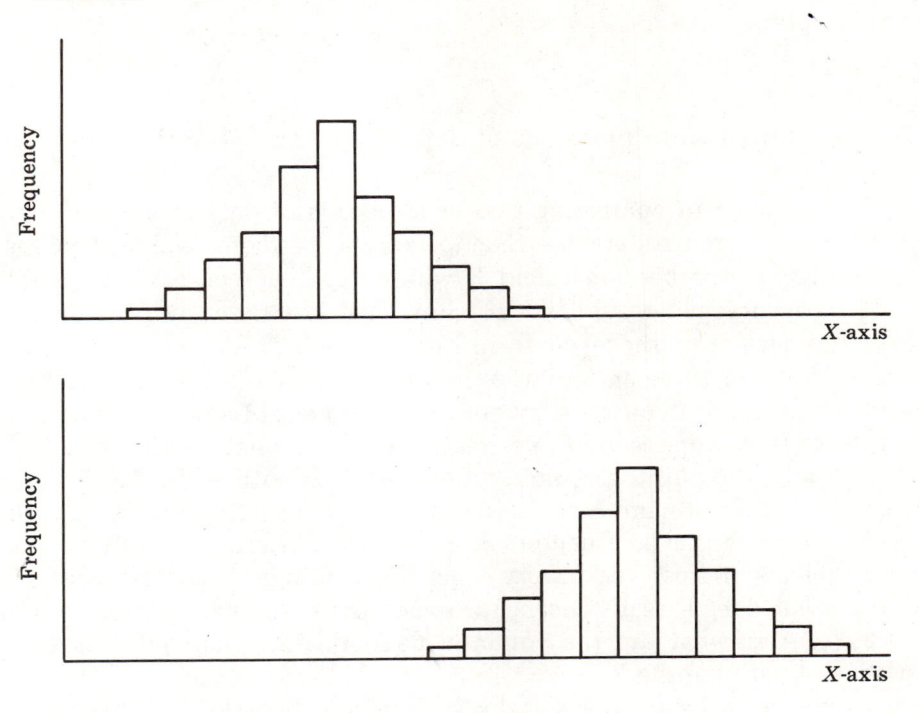

3. How *dispersed*, or spread out, the observations are.

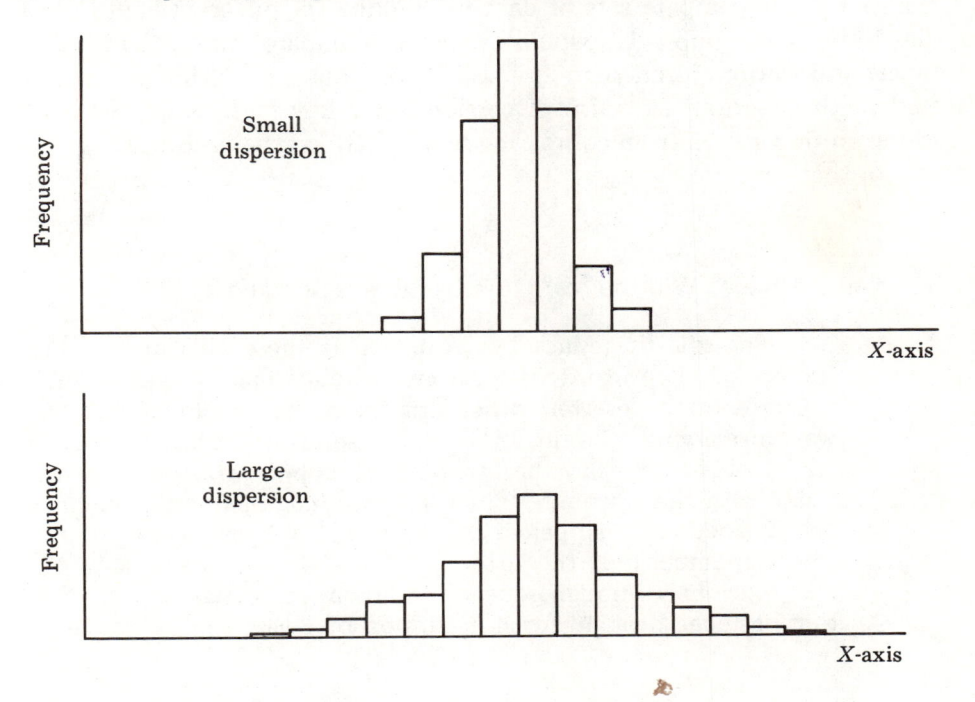

All the methods and procedures, both graphical and numerical, for answering (1), (2), and (3) are known collectively as *descriptive statistics*. Many are illustrated in this chapter.

1.2 Comparing two or more sets of data (Examples 1.1-1.3)

The problem of comparing two or more sets of data is one of the most frequently encountered in statistics. Examples are everywhere. The medical researcher wants to know whether Treatment X will have a higher cure rate than Treatment Y. The psychologist would like to show that rats raised in Environment 1 are more aggressive than those raised in Environment 2. The engineer needs to determine which of three antipollution devices is most effective. The simple fact is that it would be difficult to carry out an experiment of any consequence that did *not* call for the comparison of two treatments, at one point or another.

Actually, the purpose of this section is not to outline the formal procedures for comparing Treatment A to Treatment B or correlating Factor X with Factor Y. We will leave that to Chapters 3, 4, 5, and 6. Here we simply want to introduce problems of this type in very general terms, to get an overview of what they entail. What does it really mean, for example, to compare one set of data with another? And what are the common denominators that link one experiment statistically with another?

As you read the examples in this section, note how the word "compare" takes on different meanings in different contexts. Sometimes, as in the case of Examples 1.1 and 1.2, we compare sets of data by looking for differences in their shapes. Other times, as Example 1.3 would indicate, "compare" means to check for differences in location. And there are still other situations (although none are illustrated in this section) in which dispersion is the key factor. Of course, there is nothing to prevent us from comparing sets of data on the basis of two, or even all three, of these criteria.

Example 1.1   "Will the *real* Quintus Curtius Snodgrass..."

The role that Mark Twain played in the Civil War has long been a subject of historical debate. Some scholars have suggested that he was a Confederate deserter; others maintain that his loyalty to the South was unswerving. Twain, himself, refused to shed any light on the matter. In his later years, he put off biographers who pressed him on the point with the warning, "When I was younger, I could remember anything, whether it happened or not; but now I am getting old and soon I shall remember only the latter."

Curiously enough, one of the more important pieces of evidence that has been brought forth in support of Twain's military achievements is a set of ten letters that appeared in the *New Orleans Daily Crescent* during 1861. Signed "Quintus Curtius Snodgrass," the letters purported to chronicle the author's adventures as a member of the Louisiana militia. While historians generally agree that the accounts referred to actually *did* happen, there seems to be no record of anyone named Quintus Curtius Snodgrass. Adding to the mystery is the fact that the style of the letters bears unmistakable traces—at least to some critics—of the humor and irony that made Mark Twain so famous.

In the past, incidents of disputed authorship have not been uncommon. Speculation has persisted for several hundred years that some of Shakespeare's works were written by Sir Francis Bacon. And whether it was Alexander Hamilton or James Madison who wrote certain of the Federalist Papers is still an open question. The analysis in this example shows how descriptive statistics can be used to solve what appears to be a totally nonstatistical problem—in a totally nonstatistical subject.

Objective

To determine whether Mark Twain could have been the real author of the Quintus Curtius Snodgrass letters.

Procedure

Efforts to unravel these "yes, he did—no, he didn't" controversies often rely heavily on literary and historical clues. But not always. Studies have shown that authors, like burglars, leave "fingerprints"—only those of the former are verbal. A given author will use roughly the same proportion of, say, three-letter words in something he writes this year as he did in whatever he wrote last year. The same holds true for words of any length. *But*, the proportion of three-letter words that Author A consistently uses will very likely be different than the proportion of three-letter words that Author B uses. Theoretically, then, by constructing a word-length frequency count for essays known to be written by Mark Twain and comparing that to a similar count made for the Snodgrass letters, it should be possible to assess the likelihood of the two authors being one and the same.

Data

Table 1 and Figure 1 show the word-length distributions for three sets of letters known to have been written by Mark Twain. (Of the 1,885 words making up Sample 1, 74 were one-letter words, 349 were two-letter words, and so on.

Table 1
Word-length Distributions for Three Samples of Mark Twain's Writings

Word Length	Sample 1 Frequency	Sample 1 Relative Frequency	Sample 2 Frequency	Sample 2 Relative Frequency	Sample 3 Frequency	Sample 3 Relative Frequency
1	74	.039	312	.051	116	.039
2	349	.185	1,146	.188	496	.167
3	456	.242	1,394	.228	673	.226
4	374	.198	1,177	.193	565	.190
5	212	.113	661	.108	381	.128
6	127	.067	442	.072	249	.084
7	107	.057	367	.060	185	.062
8	84	.045	231	.038	125	.042
9	45	.024	181	.030	94	.032
10	27	.014	109	.018	51	.017
11	13	.007	50	.008	23	.008
12	8	.004	24	.004	8	.003
13+	9	.005	12	.002	8	.005
	1,885		6,106		2,974	

Figure 1

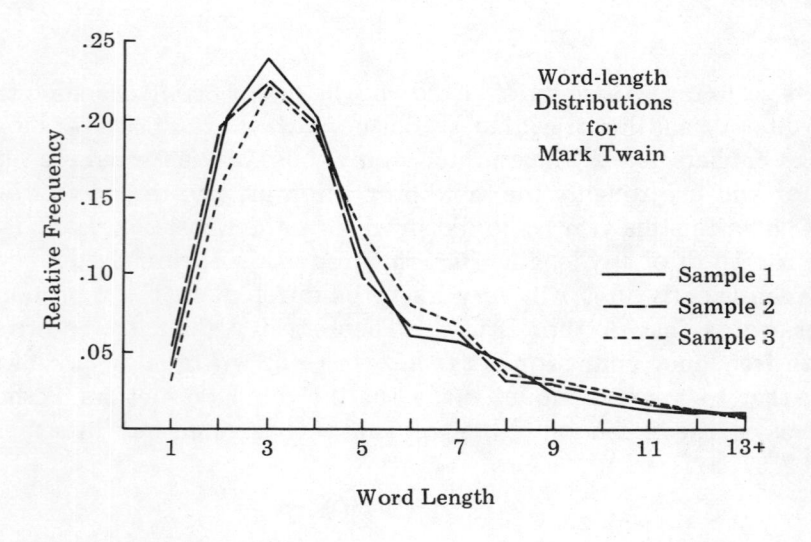

Word-length Distributions for Mark Twain

— Sample 1
— — Sample 2
- - - - Sample 3

Question 1.1.1

Why is the vertical axis of Figure 1 scaled in terms of relative frequencies rather than the raw frequencies listed in Table 1?

Question 1.1.2

On the basis of the three relative frequency polygons of Figure 1, what would you conclude about the consistency of Twain's word-length distributions?

The next table gives the combined word-length counts (and relative frequencies) for the ten Snodgrass letters.

Table 2
Word-length Distribution for the Ten QCS Letters

Word Length	Frequency	Relative Frequency
1	424	.032
2	2,685	.204
3	2,752	.209
4	2,302	.175
5	1,431	.109
6	992	.075
7	896	.068
8	638	.048
9	465	.035
10	276	.021
11	152	.011
12	101	.008
13+	61	.005
	13,175	1.000

Question 1.1.3

Superimpose a relative frequency polygon for the Snodgrass distribution over the three Twain distributions graphed in Figure 1. What is your conclusion?

Question 1.1.4

Could an authorship dispute be resolved by taking *one* sample from each of the authors in question and comparing the two word-length distributions?

Question 1.1.5

Questions 1.1.2 and 1.1.3 referred to the *shapes* of the Twain distributions and the QCS distribution. How might the *locations* of these distributions be determined?

Question 1.1.6

Draw a relative cumulative frequency polygon for the Snodgrass distribution and graphically estimate the 50th percentile.

Question 1.1.7

Would the 50th percentile obtained in Question 1.1.6 numerically equal the mean of the QCS distributions? Explain. (Don't do any computations).

Question 1.1.8

In making frequency counts like the ones shown in Tables 1 and 2, what kinds of words, if any, do you think should be excluded?

Question 1.1.9

In a paternity suit, certain medical information (blood type, length of pregnancy, and so on) can sometimes be introduced that will prove the defendant is *not* the child's father. But there is no way that such information can prove that the defendant *is* the child's father. Is it likely that statistical tests of authorship are similarly "one sided"? Explain.

REFERENCES

1. Brinegar, Claude S., "Mark Twain and the Quintus Curtius Snodgrass Letters: A Statistical Test of Authorship" (*Journal of the American Statistical Association*, 58, 1963, pp. 85-96).
2. Mosteller, Frederick, and David Wallace, "Inference in an Authorship Problem" (*Journal of the American Statistical Association*, 58, 1963, pp. 275-309).
3. Williams, C. B., "Studies in the History of Probability and Statistics" (*Biometrika*, 43, 1956, pp. 248-256).

Example 1.2 Politics and academia

Until recently, very little was known about the political leanings of university faculties. It was tacitly assumed that a teacher would vote according to the patterns established by his or her ethnic group, religious affiliation, or socioeconomic level. According to the following study, though, that may have been an oversimplification. In 1969 an extensive survey of American college and university professors revealed that their political preferences were closely tied to their fields of interest. A typical physicist, for example, will react to public issues quite differently than a typical engineer, even though the two might be next-door neighbors and identical with respect to all the usual political indices.

The political philosophies of faculty members from four different academic disciplines—agriculture, medicine, law, and social science—are summarized in this example. Note that this is precisely the sort of situation in which the frequency polygon becomes a most effective graphical technique. If we were to draw three or more histograms on the same graph the amount of overlap from histogram to histogram would most probably render the entire graph unintelligible. With the frequency polygon format, though, the problem of overlap is never a factor.

Objective

To characterize and contrast the political preferences of college and university professors.

Procedure

A total of some 60,000 full-time faculty members responded to the survey. From the almost 300 questions asked, each person was classified on a five-point scale, as being either (1) very liberal, (2) liberal, (3) middle-of-the-road, (4) conservative or, (5) very conservative.

Pigeonholing all the respondents into just five classes is not easy. Many individuals are going to take liberal positions on some issues and conservative positions on others. Ultimately, classifications of this sort are done using a statistical technique known as factor analysis. Each question is thought of as a "factor" capable of discriminating one political philosophy from another. Some questions, of course, prove to be better discriminators than others. In this particular survey, four issues emerged as being central to the split between liberals and conservatives: (1) the Vietnam war, (2) legalization of marihuana, (3) the causes of black riots, and (4) busing as a means for school integration. (How these four questions were singled out and the way the final classification was made will not be considered here. For further details, consult Ludd and Lipset (1972).)

Data

Figure 1 details the political philosophy "pattern" for respondents from each of the four disciplines in question. It was necessary to scale the vertical axis as a percentage because the numbers in each of the groups were different.

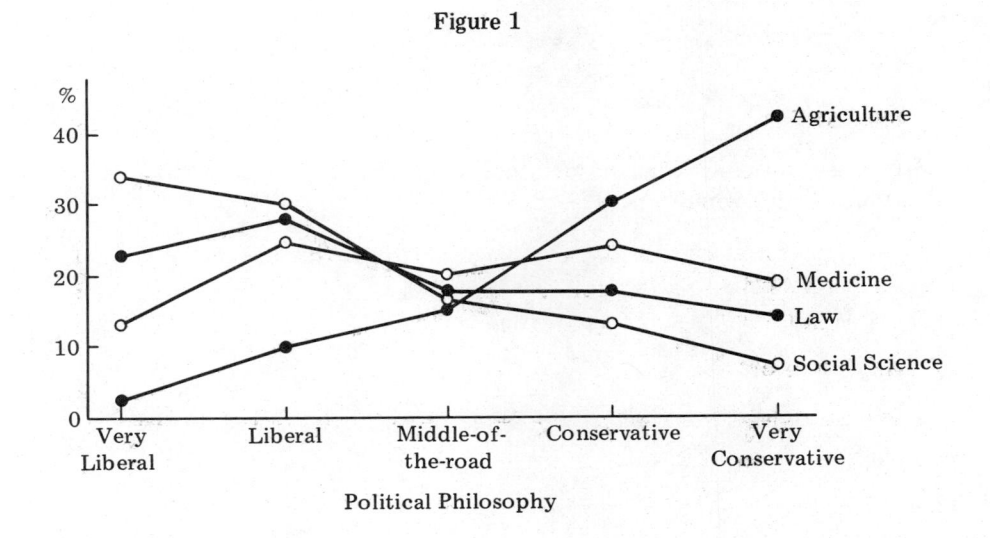

Figure 1

Question 1.2.1

Since this survey was done by mail, the respondents no longer represented a random sample but, rather, a sample of all those who, if solicited by mail, would reply. How might this affect the results?

Question 1.2.2

What kind of scale—nominal, ordinal, or interval—do these data represent?

Question 1.2.3

The sample mean of n observations $x_1, x_2, \ldots, x_n$ is given by the formula

$$\overline{x} = (1/n) \sum_{i=1}^{n} x_i$$

Suppose, though, that associated with the ith observation is a certain "weight," w_i, where

$$\sum_{i=1}^{n} w_i = 1.$$

Then we define the *weighted average* of $x_1, x_2, \ldots, x_n$ to be

$$\overline{x}_w = \sum_{i=1}^{n} w_i x_i.$$

In Figure 1, replace the descriptions "very liberal," "liberal," ..., "very conservative" with a numerical scale, "0," "1," ..., "4," and find weighted averages for each of the four academic disciplines.

Question 1.2.4

What assumption is being made when the original scale of Figure 1 is replaced by the numbers 0 through 4? How do we know that the substitution should not be −1, 2, 3½, 7, 11 instead of 0, 1, 2, 3, 4?

Question 1.2.5

What sort of response pattern do you think would be characteristic of engineering professors?

REFERENCE

1. Ludd, Everett Carll, Jr., and Seymore M. Lipset, "Politics of Academic Natural Scientists and Engineers" (*Science*, 176, 1972, pp. 1091-1100).

Example 1.3 Etruscan skulls

In the eighth century B.C., the Etruscan civilization was the most advanced in all of Italy. Its art forms and political innovations were destined to leave indelible marks on the entire Western world. Originally located along the western coast between the Arno and Tiber Rivers (the region now known as Tuscany), it spread quickly across the Apennines and eventually overran much of Italy. But as quickly as it came, it faded. Militarily it was to prove no match for the burgeoning Roman legions, and by the dawn of Christianity it was all but gone.

No chronicles of the Etruscan empire have ever been found, and to this day its origins remain shrouded in mystery. Were the Etruscans native Italians or were they immigrants? And if they were immigrants, where did they come from? Much of what *is* known has come from archeological investigations and anthropometric studies—the latter involving the use of body measurements to determine racial characteristics and ethnic origins. The data presented here are an example of one such study.

Objective

To test the hypothesis that the Etruscans were native to Italy by comparing skull measurements made on present-day Italians with those made on specimens known to be the remains of Etruscans.

Data

Table 1 lists the maximum head breadths (in mm) found for 84 male Etruscans. Table 2 gives the same information for 70 modern Italians.

Table 1
Maximum Head Breadths, in mm, of 84 Etruscan Males

141	148	132	138	154	142	150
146	155	158	150	140	147	148
144	150	149	145	149	158	143
141	144	144	126	140	144	142
141	140	145	135	147	146	141
136	140	146	142	137	148	154
137	139	143	140	131	143	141
149	148	135	148	152	143	144
141	143	147	146	150	132	142
142	143	153	149	146	149	138
142	149	142	137	134	144	146
147	140	142	140	137	152	145

Table 2
Maximum Head Breadths, in mm, of 70 Modern Italian Males

133	138	130	138	134	127	128
138	136	131	126	120	124	132
132	125	139	127	133	136	121
131	125	130	129	125	136	131
132	127	129	132	116	134	125
128	139	132	130	132	128	139
135	133	128	130	130	143	144
137	140	136	135	126	139	131
133	138	133	137	140	130	137
134	130	148	135	138	135	138

Question 1.3.1

In the spaces that follow construct relative frequency distributions for the data of Tables 1 and 2. Also, draw a histogram for each distribution. Put both histograms on the same graph.

Table 3
Etruscan Skull Measurements

Maximum Breadth (mm)	Frequency	Relative Frequency

Table 4
Italian Skull Measurements

Maximum Breadth (mm)	Frequency	Relative Frequency

Question 1.3.2

On the basis of the two histograms drawn for Question 1.3.1, what tentative conclusion would you reach about the "Italian" theory? Which characteristic of variability—shape, location, or dispersion—had the most influence on your decision? What other data would you want to examine before taking a firm stand?

Question 1.3.3

Compute the ungrouped mean, $\bar{x}$, and the grouped mean, $\bar{x}_g$, for the data of Tables 2 and 4. In general, is $\bar{x}_g$ as likely to be smaller than $\bar{x}$ as it is to be larger than $\bar{x}$?

Question 1.3.4

Compute the ungrouped standard deviation, s, and the grouped standard deviation, s_g, for the data of Tables 2 and 4.

Question 1.3.5

Draw a cumulative frequency polygon for the data of Table 1. Graphically estimate the *interquartile range*, the difference between the 75th and 25th percentiles. (For distributions not having a bell shape, the interquartile range is sometimes used in place of the standard deviation as a measure of dispersion.)

Question 1.3.6

For most data, the range (largest observation - smallest observation) will be from three to five times as large as the standard deviation (either s or s_g.) Is this true for the data of Tables 1 and 2? Note that this relationship provides a very quick computational check for s.

REFERENCES

1. Barnicot, N. A., and D. R. Brothwell, "The Evaluation of Metrical Data in the Comparison of Ancient and Modern Bones" (in *Medical Biology and Etruscan Origins*, G. E. W. Wolstenholme and Cecilia M. O'Connor, eds., Little, Brown, and Co., 1959, p. 136).
2. Fell, R. A. L., *Etruria and Rome* (Cambridge University Press, 1924, pp. 1-82).
3. Randall-Maciver, David, *The Etruscans* (Oxford University Press, 1927, pp. 1-18).

1.3 Characterizing a Single Set of Data (Examples 1.4-1.6)

The characterization of variability is seldom the final objective of an experiment yet it always plays a major role in reaching that objective. There is no way, for example, that comparisons of the kind described in the previous section can be made until the variability of the measured responses has been determined. In a very real sense, variability acts as a yardstick against which the apparent treatment effects are scaled.

In mathematical terms, variability is defined in terms of a quantity known as the *sample standard deviation*, s. Given a set of n observations $x_1, x_2, \ldots, x_n$.

$$s = \sqrt{\frac{\sum_{i=1}^{n}(x_i - \bar{x})^2}{n-1}}$$

where $\bar{x}$ is the sample mean

$$\bar{x} = (1/n)\sum_{i=1}^{n} x_i.$$

In words, we would say that s is the square root of the average of the squared deviations from $\bar{x}$.

The standard deviation measures the extent to which the sample observations deviate from one another. If all the x_i's were equal, $\bar{x}$ would equal x_i, each $(x_i - \bar{x})^2$ term would equal 0, and s would be 0. By the same token, as the x_i's become more and more dispersed, the $(x_i - \bar{x})^2$ terms become larger and larger, and the standard deviation increases. (Why is it necessary to square the deviations and then take the square root of their sum? What would be wrong with using

$$\frac{\sum_{i=1}^{n} (x_i - \bar{x})}{n-1}$$

as a measure of dispersion?)

We call the expression

$$\sqrt{\frac{\sum_{i=1}^{n} (x_i - \bar{x})^2}{n-1}}$$

the *defining formula for s*. Unfortunately, the computations it calls for are rather cumbersome. The term $\sum_{i=1}^{n} (x_i - \bar{x})^2$, for example, is quite tedious to evaluate, even with a desk calculator. For this reason, we will always use another formula for s when we work problems:

$$s = \sqrt{\frac{n \sum_{i=1}^{n} x_i^2 - (\sum_{i=1}^{n} x_i)^2}{n(n-1)}}$$

This is called the *computing formula for s*. It may not look it, but this is really much easier to work with than the defining formula. Algebraically the two are equivalent: for any set of data

$$\sqrt{\frac{\sum_{i=1}^{n} (x_i - \bar{x})^2}{n-1}} = \sqrt{\frac{n \sum_{i=1}^{n} x_i^2 - (\sum_{i=1}^{n} x_i)^2}{n(n-1)}}$$

Keep in mind that the words standard deviation and variability are not synonymous. The standard deviation is simply a mathematical formula for quantifying one particular aspect of variability. But variability, itself, is a much broader concept and refers in a very general way to the way in which the observations differ among themselves. In fact, in the examples presented in this section, we will characterize variability in terms of the shape and location of the observations as well as their standard deviation.

Example 1.4 Fingerprints

As an illustration of what statisticians mean by the concept of variability, there is probably no more familiar an example than the fingerprint. Formed during the beginning of the second trimester, the delicate swirls and ridges that make up a fingerprint leave each individual with a unique identity. No two are the same (even in "identical" twins), and their original pattern never changes.

But despite their manifest differences, fingerprints do have certain things in common. Toward the end of the last century, Sir Francis Galton was able to classify all fingerprints into three generic types: the whorl, the loop, and the arch.

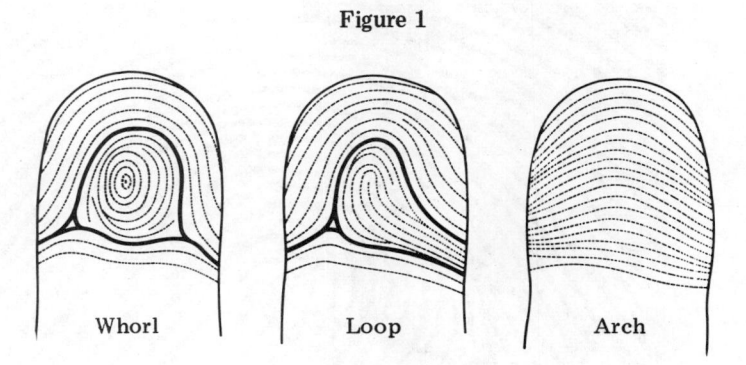

Figure 1

Whorl   Loop   Arch

A few years later, Sir Edward Richard Henry, a pioneer criminologist who was later to become Commissioner of Scotland Yard, refined Galton's system to include eight generic types. The Henry system, as it came to be known, revolutionized the problem of criminal identification and is still used today by the FBI.

Not unexpectedly, the classification of fingerprints gave rise to a form of fortune telling (known more elegantly as dactylomancy.) According to those who believe in such things, a person "having whorls on all fingers is restless, vacillating, doubting, sensitive, clever, eager for action, and inclined to crime." Likewise, a "mixture of loops and whorls signifies a neutral character, a person who is kind, obedient, truthful, but often undecided and impatient" (see Cummins and Midlo).

Objective

To characterize in a quantitative way the variability associated with fingerprints.

Procedure

There are many characteristics besides the three proposed by Galton and the eight proposed by Henry that can be used to distinguish one fingerprint from another. One of these is the *ridge count*. To see how the ridge count is defined, look again at Figure 1. In the loop pattern there is a point where the three opposing ridge systems come together. This is known as the triradius. If a straight line is drawn from the triradius to the center of the loop, a certain number of ridges will be crossed. In Figure 2 that number is 11.

Figure 2

If the number of crossings is determined for each finger and the results added together, the resultant sum is known as the ridge count.

Data

The following table shows the distribution of ridge counts found for 825 males.

Table 1
Ridge Counts

Count	Frequency
0- 19	10
20- 39	12
40- 59	24
60- 79	40
80- 99	73
100-119	100
120-139	90
140-159	117
160-179	139
180-199	100
200-219	67
220-239	36
240-259	10
260-279	4
280-299	3
	825

Question 1.4.1

Suppose a set of fingerprints was found at the scene of a crime and it was determined that the ridge count was at least 250 (the exact value being in question because of smudging). Furthermore, suppose a potential suspect has a ridge count of 261. What would you conclude and why?

Question 1.4.2

Draw a histogram for these data

Question 1.4.3

Suppose similar measurements were made on a group of 10,000 males. About how many would have ridge counts between 120 and 179, inclusive?

Question 1.4.4

Draw a cumulative frequency polygon for the data of Table 1 and estimate the 25th, 50th, and 75th percentiles.

Question 1.4.5

By virtue of their fetal origin, fingerprint characteristics are almost entirely the consequence of genetic, rather than environmental, factors. Suppose that x denotes a mother's ridge count and y, her child's. In the first graph below, draw a set of points that you think would be typical of the relationship between x and y. Do the same in the second graph for a set of points where the x- and y-coordinates are a husband's and wife's ridge counts, respectively.

Question 1.4.6

For the data of Table 1, $\bar{x}_g = 145.8$ and $s_g = 51.8$. Approximately what proportion of the observations are in the interval $(\bar{x}_g - s_g, \bar{x}_g + s_g)$? What proportion are in the interval $(\bar{x}_g - 2s_g, \bar{x}_g + 2s_g)$? (For distributions having a "perfect" bell shape, 68 percent of the observations will lie within one standard deviation of the mean, 95 percent within two standard deviations.)

REFERENCES

1. Carter, C. O., "Multifactorial Genetic Disease" (*Hospital Practice*, 5, 1970, pp. 45-59).
2. Harold Cummins and Charles Midlo, *Finger Prints, Palms, and Soles* (Blakiston Company, 1943).

Example 1.5 Making decisions

How do we make decisions? What factors do we appeal to and how do the influences of those factors vary from person to person?

In this example we try to answer those questions by examining a model for the decision-making process.

Let's assume—even at the risk of oversimplification—that all decision-making motivations can be put under two broad headings: "legitimacy" factors and "sanction" factors. The former refer to motivations arising out of feelings the decision maker has of what is "right." Persons with this orientation are called *moralists*. Sanction factors, on the other hand, refer to influences like peer-group pressure, short-range gain, the desire to be liked, and so forth. Persons giving heavy consideration to these sorts of factors are called *expedients*. Of course, in addition to these two extremes, there are people who sometimes act out of righteousness and other times out of expediency. The question is this: how is the total population of potential decision makers likely to be distributed along this particular rationale scale? Notice in this example how a numerical structure is superimposed over what is initially a nonnumerical problem.

Objective

To study the distribution of motivations that people have when they make decisions.

Procedure

A group of 106 subjects were given a test that consisted of 37 different conflict situations. A possible solution was proposed for each situation. The subjects had to decide—if it was up to them—whether *they* would resolve the conflict in the manner that was suggested. But instead of answering just "yes" or "no," they were required to indicate the strength of their conviction in taking (or not taking) the recommended course of action. Their possible responses were:

1. Absolutely must (take the action).
2. Preferably should (take the action).
3. May or may not (take the action).
4. Preferably should not (take the action).
5. Absolutely must not (take the action).

It was felt that the "absolutely must" and "absolutely must not" responses were probably moralistic in origin and a subject was given a "point" each time he gave either one. The total number of points accumulated over the 37 questions was taken as an index of the extent to which a subject sought moral solutions.

Data

The scores of the 106 subjects are listed in Table 1.

Table 1
Motivation Scores

13	12	11	19	24	2	13
17	15	2	17	15	7	15
13	27	4	16	13	9	5
8	19	4	17	12	5	28
7	23	13	13	6	21	20
10	6	10	7	17	18	19
10	2	13	9	27	17	14
21	9	19	12	3	18	11
18	11	25	11	10	12	14
17	5	14	30	7	15	4
19	18	11	19	1	13	8
15	20	4	4	14	13	10
15	24	14	11	22	15	7
23	15	12	18	16	6	23
12	14	23	18	10	25	18
						24

Question 1.5.1

Make up a frequency distribution for these data and draw the corresponding histogram in the spaces provided on the next page.

Table 2

Score	Frequency

Question 1.5.2

How might a personnel department make use of this information? What sort of jobs would best be filled by a moralist? By an expedient?

Question 1.5.3

Compute $\bar{x}_g$ using the classes defined in Table 2.

Score	Midpoint	Frequency	Midpoint × Frequency

Question 1.5.4

Suppose that you hire a night watchman who claimed he had taken this test somewhere else and gotten a score of 29. A week later, you catch him stealing. What would you conclude? Suppose he claimed his score was 16. How would that affect your conclusion?

Question 1.5.5

Estimate the chances that a randomly selected person would have a score between 19 and 21, inclusive.

Question 1.5.6

The first six columns in Table 1 contain 15 scores, the seventh column, 16 scores. Subdivide each column into three groups of five numbers: 13 through 7, 10 through 17, and 19 through 12 for the first column, and similarly for the other columns. Omit the 106th observation (24) in Column 7. Compute the mean for each of the 21-column subdivisions. Make a histogram for these 21 sample means, using the same scale as in Question 1.5.1. How would you compare the distribution of the original scores with the distribution of the average scores?

REFERENCE

1. Gross, Neal, Ward S. Mason, and Alexander W. McEachern, *Explorations in Role Analysis* (Wiley, 1958, p. 297).

Example 1.6 Bird calls

Bird calls are surprisingly difficult to distinguish. It would be easy if, say, all blue jays sang one song and all robins another. Unfortunately, that isn't what happens. For one thing, birds have dialects: members of a given species living in one area sing differently than birds of the same species living in a different area. Even more of a problem, though, is the fact that birds are not limited to a single song—they have an entire repertoire.

The basic building block of all bird songs is the *syllable*. A bird such as the cardinal can make at least ten distinct sounds, or syllables. Individual syllables sung in rapid succession are said to belong to the same *utterance*, and a series of consecutive utterances is called a *bout*.

Defining a bout in this way—as a series of utterances—raises obvious statistical questions. How many utterances, for example, typically make up a bout? And how does that number vary from bout to bout?

Objective

To characterize the variability in the number of utterances per bout for the North American cardinal.

Procedure

The subjects of this study were male cardinals (*Richmondena cardinalis*) nesting near the campus of the University of Western Ontario. Over a period of several months, the songs of these birds were taped and analyzed with a sound spectrograph. This is a device that graphs the frequency of a bird's call as a function of time. Spectrographs of two of the syllables in a cardinal's repertoire are shown here. The first is known as Song Type D; the second, Song Type Q.

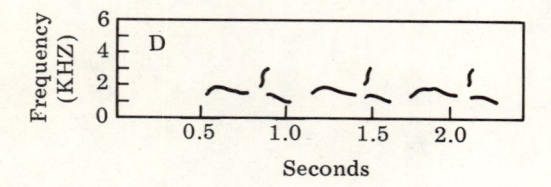

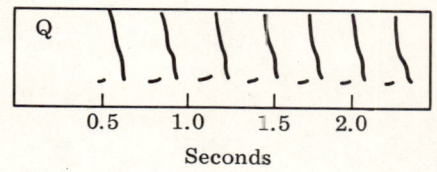

Any attempt to characterize statistically the song patterns of cardinals has to look at many different kinds of questions. How long, for example, is a typical utterance made up of Q syllables? And does that average length differ from the average length of utterances composed of D syllables? When a cardinal combines two or more syllables in a single utterance, do certain syllables tend to follow certain other ones? Or are the patterns random? Here we look at still another variable, the number of utterances included in a single bout.

Data

Table 1 and Figure 1 show the number of utterances in bouts of Song Type D. A total of 250 bouts were recorded.

Table 1
Distribution of the Number of Utterances in Bouts of Song Type D

Number of Utterances per Bout	Frequency	Relative Frequency
1	132	.528
2	52	.208
3	34	.136
4	9	.036
5	7	.028
6	5	.020
7	5	.020
8+	6	.024
	250	1.00

Figure 1

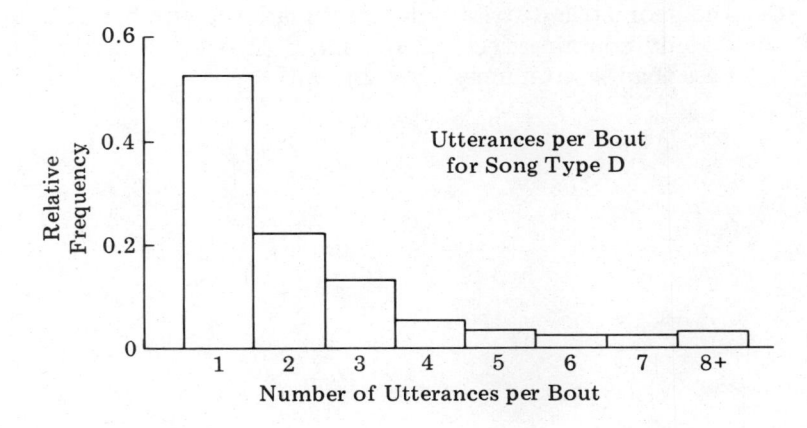

Utterances per Bout for Song Type D

Analysis

In describing the variability characteristic of a given phenomenon, it often helps to introduce the notion of a *mathematical model*. This is simply a formula that can predict with reasonable accuracy the observed distribution of the measured responses. For this experiment, the response being measured is X, the number of utterances in a single bout. This means that a mathematical model for X would have to be capable of predicting the number of times—out of, say, N bouts—that X will equal 1, how many times it will equal 2, and so on.

For reasons that cannot be gone into here, one formula that might serve as a mathematical model for X is the following:

$$\text{expected number of times that } X \text{ will equal } k = N\left(\frac{1}{\bar{x}}\right)\left(1 - \frac{1}{\bar{x}}\right)^{k-1}, \text{ for } k = 1, 2, \ldots$$

where $\bar{x}$ is the average number of utterances per bout. This particular mathematical model is known as the *geometric distribution* (and X is said to be a geometric *random variable*).

Question 1.6.1

Compute $\bar{x}$ for the data in Table 1. Assume that each of the 6 bouts having "8+" utterances had, in fact, 8 utterances.

Question 1.6.2

Use the geometric distribution to estimate the number of bouts (out of 250) that would contain exactly 1 utterance ($X = 1$). Exactly 2 utterances ($X = 2$). Less than 4 utterances ($X < 4$).

Question 1.6.3

Suppose that we wanted to measure the agreement between the observed and expected frequencies (as a means of determining the appropriateness of the presumed model). Would it be feasible to compute the algebraic difference between the observed and expected frequencies for a given value of X and then add those differences for all the values of X? That is, can "goodness-of-fit" be measured by the expression

$$\sum_{\text{all values of } X} (\text{observed frequency} - \text{expected frequency}) ?$$

Explain. (Note: the sum of the expected frequencies will always be equal to the sum of the observed frequencies—in this case, 250.)

Question 1.6.4

Examples of mathematical models are quite common in genetics. The feathers of a particular kind of chicken known as a frizzle fowl come in three variations (or phenotypes)—extreme frizzle, mild frizzle, and normal. The effect of the two alleles (F and f) that control frizzle is an example of a phenomenon known as incomplete dominance. Specifically, progeny whose genetic complement is (F, F) are extreme frizzles, those having an (F, f) genotype are mild frizzles, while the (f, f) offspring are normal. In crosses between mild frizzle parents, a total of 93 first-generation hybrids were produced:

Phenotype	Observed Frequency
Extreme frizzle	23
Mild frizzle	50
Normal	20
	93

What would be the obvious mathematical model in this instance? Use the model to calculate expected frequencies for each of the three phenotypes.

REFERENCE

1. Lemon, Robert E., and Christopher Chatfield, "Organization of Song in Cardinals" (*Animal Behaviour*, 19, 1971, pp. 1–17).

1.4 Interpreting the Unexpected (Example 1.7)

Sometimes we set out to describe what appears to be a very routine phenomenon only to end up with much more than we bargained for. Example 1.7 is a case in point. In that particular study, the shape of the response distribution was very peculiar and, in so being, revealed the presence of factors that were totally unanticipated when the experiment was first begun. The consequence was that the interpretation of the results was turned completely around.

If there is such a thing as a "typical" response pattern, it would have to be the bell-shaped distribution, like the ones in Examples 1.3, 1.4, and 1.5. Whenever many factors are contributing to the numerical value of the response variable, the resulting histogram often has this shape. We know, for instance, that the width of an Etruscan's skull would be determined by many different genetic and environmental factors. It comes as no surprise, then, that the distribution of widths is bell shaped (see Example 1.3).

When a response pattern is very "nonbell shaped"—say, bimodal or U-shaped—it is often an indication that a limited number of factors are dominating all the others. This is what happened in Example 1.7. Learning what those factors are then becomes the primary objective of the analysis.

As you read descriptions of experiments, get into the habit of predicting what the response distribution will look like and what its location and dispersion will be. Only by knowing the expected can we profit from the unexpected.

Example 1.7 Comparing headache remedies

Some years ago the manufacturers of a certain headache remedy thought it might be a good idea to see whether all the ingredients they were putting into their product were really necessary. They decided to test four different preparations (A, B, C, and D), all identical from outward appearances. Those designated "A" were the same product they were currently marketing; "B" and "C" had some, but not all, of the presumed active ingredients of "A"; "D" was a placebo—it contained nothing that physiologically could alleviate the symptoms of a headache.

From a statistical standpoint, the problem seemed simple enough. All that was necessary was to find four groups of subjects, see what proportion of the time each preparation was successful in relieving headaches, and draw the (one hoped) obvious conclusion. But things didn't turn out quite the way they were supposed to . . . .

> **Objective**
>
> To compare the effectiveness of four different headache remedies.

Procedure

A total of 199 persons, all complaining of frequent headaches, agreed to participate in the study, which was scheduled to last eight weeks. They were divided into four groups (I, II, III, and IV). For two weeks, each group was given a supply of one of the four preparations. Any time a subject had a headache during that period, he or she was instructed to take the assigned drug and, after waiting 30 minutes, record whether it worked. When the two weeks were up the drugs were switched, so that after eight weeks each person had used each treatment. (At no time, of course, were the identities of the preparations revealed to the subjects.) The experimental "design"—that is, the way in which the treatments were allocated to the subjects—is shown in Table 1.

Table 1
Treatment Allocations

Group	1st Two Weeks	2nd Two Weeks	3rd Two Weeks	4th Two Weeks
I	A	B	C	D
II	B	A	D	C
III	C	D	A	B
IV	D	C	B	A

Question 1.7.1

How would you describe the allocation of the four headache remedies among "groups" and "weeks"? (This particular experimental design is known as a *Latin square* (see Question 1.7.3).)

When the eight weeks were up, it was an easy task to compute the overall proportion of headaches that each of the treatments cured. But the results, given in Table 2, came as quite a surprise.

Table 2
Per cent of Headaches Relieved

A	B	C	D
84	80	80	52

It was not unexpected that Preparation A proved to be the most effective (it had more active ingredients than the other three), but the fact that the placebo was successful 52 per cent of the time was disturbing. A closer look at the data was clearly in order.

Data

At this point in a statistical analysis there is no obvious route to follow. All we know for sure is that something totally unanticipated has happened: in this case, the high success rate for the placebo. What matters now is to explain *why* it happened and how it will affect the interpretation of the rest of the experiment. Of course, this puts the statistician in the uncomfortable position of having to find something without knowing what it is he is looking for.

One hopes that, by examining the data from many different perspectives, some sort of pattern will emerge. Here everything fell into place when the responses of the placebo group were considered in more detail. Table 3 lists the number of "cures" reported among the 59 persons who were using the placebo and who complained of exactly five headaches during that particular two-week period.

Table 3
Number of Headaches Relieved
(Out of Five) by the Placebo

0	4	5	4	0	5
0	0	2	0	2	3
2	0	5	3	2	0
3	5	0	0	4	0
0	5	0	5	5	4
4	2	0	4	0	5
5	0	1	3	5	3
0	5	3	5	0	5
5	3	5	0	0	0
4	0	4	0	5	

Question 1.7.2

Draw a histogram for the data of Table 3. What would you conclude from the histogram and how would it modify your conduct of an experiment of this sort? Which characteristic of variability is the most revealing in this particular set of data?

Question 1.7.3

Why would the Latin square design of Figure 1 be better than the five treatment allocations indicated below? In each instance, characterize the way the treatments have been allocated (see Question 1.7.1).

Group	1st Two Weeks	2nd Two Weeks	3rd Two Weeks	4th Two Weeks
I	A	D	C	B
II	A	D	C	B
III	A	D	C	B
IV	A	D	C	B

Group	1st Two Weeks	2nd Two Weeks	3rd Two Weeks	4th Two Weeks
I	C	A	D	B
II	A	B	B	C
III	B	D	A	D
IV	B	A	C	D

Group	1st Two Weeks	2nd Two Weeks	3rd Two Weeks	4th Two Weeks
I	D	B	A	C
II	B	D	C	A
III	A	B	D	C
IV	B	A	C	D

Group	1st Two Weeks	2nd Two Weeks	3rd Two Weeks	4th Two Weeks
I	B	A	B	C
II	C	D	C	B
III	A	B	D	A
IV	D	C	A	D

Group	1st Two Weeks	2nd Two Weeks	3rd Two Weeks	4th Two Weeks
I	D	A	C	B
II	B	D	A	C
III	A	B	C	D
IV	C	D	B	A

Question 1.7.4

Suppose you decided not to use the Latin square design as described in Table 1. Which of the five designs listed in Question 1.7.3 seems like the best alternative? Why?

REFERENCE

1. Jellinek, E. M., "Clinical Tests on Comparative Effectiveness of Analgesic Drugs" (*Biometrics*, 2, pp. 87–91).

We do not what we ought,
What we ought not, we do
And lean upon the thought
That Chance will bring us through.

Arnold

Chapter 2
The One-sample Problem

2.1 Introduction

The subject of statistics is traditionally divided into two parts, *descriptive statistics* and *inference*. The emphasis of the former is on characterizing the particular sample that was actually observed; in the latter, we look beyond the sample and try to generalize about the population from which the sample was taken. Admittedly, these two areas are not perfectly distinct and they share some common ground. We saw in the last chapter, for example, that the shape and location of a histogram could often provide some very helpful insights into the population being sampled, even in the absence of any formal inference procedures. For the most part, though, the distinction is a workable one and we will follow it. Beginning in Chapter 2, we will concentrate on inference and deemphasize descriptive statistics. Chapters 2, 3, 4, and 5 profile, chapter-by-chapter, four of the most widely used "models" in inference—the one-sample problem, the two-sample problem, the paired-data problem, and the correlation problem.

The one-sample problem enjoys a unique position in statistics. In many textbooks it receives almost as much attention as all the other inference models put together. Yet in real life, the one-sample problem is not particularly common. The reason for this seemingly misplaced emphasis is twofold: (1) *all* inference procedures are structured according to the same basic principles, and (2) those principles are easiest to understand when they are developed within the framework of the one-sample problem.

The scope of the examples in Chapter 2 is broad. One is concerned with the shape of rectangles and what they reveal about a society's esthetic standards; another examines the intriguing hypothesis that people can, through an act of will, postpone dying; still another describes a behavioral syndrome that psychologists have dubbed the "Transylvania effect." On the nonhuman side, we will see how rocks are "dated," how bats find their supper, and what bees look for in flowers.

Try to anticipate, as you read the introduction to an example, the part that inference is going to play in its analysis. By familiarizing yourself with the concepts of inference as they are introduced here and by learning how to translate a verbal statement into a statistical format, much of what appears in later chapters will seem easy.

2.2 Continuous Data: the Student t Distribution (Examples 2.1-2.7)

In a one-sample problem the data consist of a single set of n measurements, each representing the same population distribution. The usual objective is to generalize (that is, make inferences) about the unknown population mean (μ) on the basis of the sample information, $x_1, x_2, \ldots, x_n$. How these inferences are made, though, depends on the *type* of data being considered. For our purposes, we will distinguish two types, *continuous* and *binomial*. Examples 2.1 through 2.7 illustrate the former; Examples 2.7 through 2.12, the latter.

There are two formats for phrasing the analysis of continuous data: hypothesis tests and confidence intervals. Mathematically the two are very similar but experimentally they arise in completely different contexts. Examples 2.1, 2.2, 2.4, and 2.5 illustrate hypothesis tests; confidence intervals are discussed in Examples 2.6 and 2.7.

In all of these problems we will be assuming, often implicitly, that the sample observations $x_1, x_2, \ldots, x_n$ represent a population distribution, P_X, that is more or less bell shaped. The question to be resolved is whether μ, the unknown mean of P_X, is equal to some specified value μ_0. Put more formally, this reduces to one of three possible hypothesis tests:

1. $H: \mu = \mu_0$
   vs.
   $A: \mu \neq \mu_0$

2. $H: \mu = \mu_0$
   vs.
   $A: \mu < \mu_0$

3. $H: \mu = \mu_0$
vs.
$A: \mu > \mu_0$

with the choice depending on the nature of the problem.

The decision to "accept H" or "reject H" is based on a procedure known as the t test. First we compute the value of

$$\frac{\bar{x} - \mu_0}{s/\sqrt{n}}$$

where $\bar{x}$ is the sample mean and s is the sample standard deviation. If P_X is bell shaped and if H is true, it can be shown that the behavior of $(\bar{x} - \mu_0)(s/\sqrt{n})$ is approximated by a Student t distribution with $n-1$ degrees of freedom. Knowing this, we can find the appropriate critical value, or values, for carrying out the test.

For example, suppose we were testing

$H: \mu = \mu_0$
vs.
$A: \mu > \mu_0$

at the $P = .05$ level of significance and suppose t^* is the number that is exceeded by t values (with $n-1$ degrees of freedom) 5% of the time.

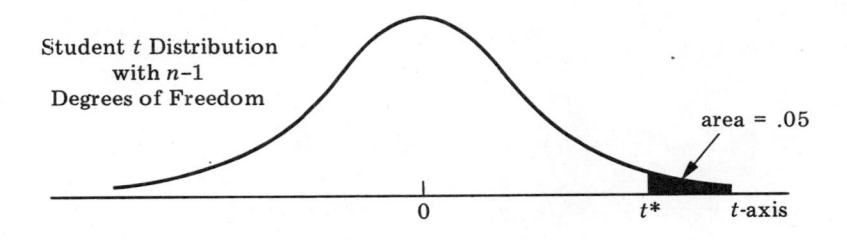

Student t Distribution with $n-1$ Degrees of Freedom

area = .05

0   t^* t-axis

Intuitively, we should reject H if $\bar{x}$ is too much larger than μ_0, or, equivalently, if

$$\frac{\bar{x} - \mu_0}{s/\sqrt{n}} \geq t^*$$

All t tests have this same basic structure. All that changes is the number and location of the critical values. For one-sided alternatives there is *one* critical value, t^*. It lies in either the left-hand tail or the right-hand tail of the appropriate

Student t distribution, its exact location being determined by the "direction" of the alternative hypothesis and the size of the level of significance. For two-sided alternatives there are *two* critical values, t_1^* and t_2^*. These are located at equal distances to the right and to the left of 0. The tail areas that t_1^* and t_2^* cut off are the same, and their sum is equal to the specified level of significance.

Confidence intervals are based on the fact that

$$\frac{\overline{X} - \mu}{s/\sqrt{n}}$$

where μ is the true mean of P_X, has a Student t distribution with $n-1$ degrees of freedom. Suppose $-t^*$ and $+t^*$ are the two values that cut off areas of .025 in either tail of that particular curve.

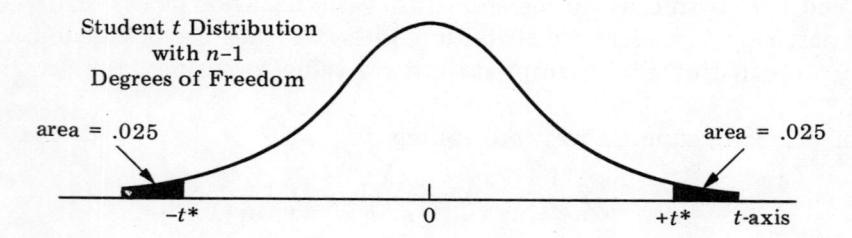

Student t Distribution with $n-1$ Degrees of Freedom

area = .025   area = .025

$-t^*$ 0 $+t^*$ t-axis

Then

$$P(-t^* < \frac{\overline{X} - \mu}{s/\sqrt{n}} < t^*) = .95$$

which can be rewritten as

$$P(\overline{X} - t^*(s/\sqrt{n}) < \mu < \overline{X} + t^*(s/\sqrt{n})) = .95$$

We call the set of values $\overline{x} - t^*(s/\sqrt{n})$ to $\overline{x} + t^*(s/\sqrt{n})$ a *95 per cent confidence interval for μ*. In the long run, 95 per cent of the intervals constructed in this fashion will contain the unknown value μ.

Example 2.1 Experimental esthetics

Not all rectangles are created equal. . . . Since early antiquity, societies have expressed esthetic preferences for rectangles having certain width (w) to length (l) ratios. Plato, for example, wrote that rectangles whose sides were in a 1: $\sqrt{3}$ ratio were especially pleasing. (These are the rectangles formed from the two halves of an equilateral triangle.)

Another "standard" calls for the width to length ratio to be equal to the ratio of the length to the sum of the width and the length. That is,

$$\frac{w}{l} = \frac{l}{w + l}$$

This implies that the width is approximately .618 times as long as the length.

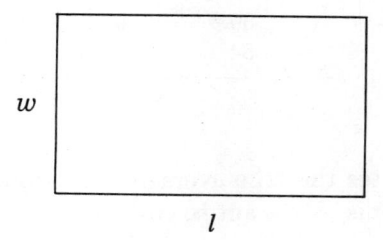

The Greeks called this the golden rectangle and used it often in their architecture. The Parthenon, for example, is dimensioned according to the golden rectangle. Other cultures also adopted this particular width to length ratio. The Egyptians built their pyramids out of stones whose faces were golden rectangles. Today, in our society, the golden rectangle remains an architectural and artistic standard. Even items like drivers' licenses, business cards, and picture frames are often made in approximately this same proportion.

The study described here is an example of a field known as experimental esthetics. The data are width-to-length ratios of beaded rectangles used by the Shoshoni Indians to decorate their leather goods. The question at issue is whether the golden rectangle can be considered an esthetic standard for the Shoshoni Indians.

Objective

To test whether the average width-to-length ratio of rectangles found on Shoshoni Indian handicraft is compatible with the Greek standard (.618).

Data

The width-to-length ratios for 20 Shoshoni rectangles are given in Table 1.

Table 1
Width-to-Length Ratios (× 100)

69.3	65.4
66.2	61.5
69.0	66.8
60.6	60.1
57.0	57.6
74.9	67.0
67.2	60.6
62.8	61.1
60.9	55.3
84.4	93.3

Analysis

Suppose μ denotes the true average width-to-length ratio, times 100, characteristic of the Shoshonis. We want to test

$$H: \quad \mu = 61.8$$

vs.

$$A: \quad \mu \neq 61.8$$

Let $P = .01$ be the level of significance. To reject H is to conclude that the Shoshonis have an esthetic standard other than the golden rectangle.

From Table 1

$$\sum_{i=1}^{20} x_i = 1321.0$$

which makes

$$\bar{x} = \frac{1321.0}{20} = 66.1$$

We need to decide whether

1. H is true and the difference between 61.8 and 66.1 is solely the result of chance

or

2. H is false and the Shoshonis do have a different standard.

Choosing between (1) and (2) requires that we know the sample-to-sample behavior of $\overline{X}$ or of some related statistic. But we do. Recall that when H is true, the sampling distribution of

$$\frac{\overline{X} - 61.8}{s/\sqrt{20}}$$

is approximated by a Student t curve with 19 degrees of freedom.

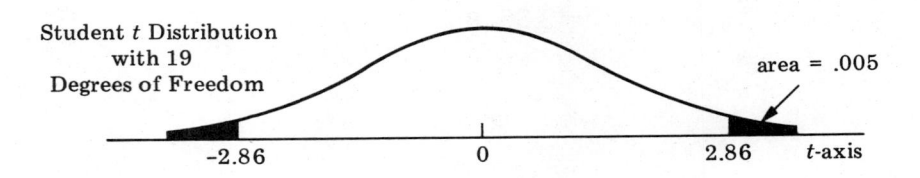

Student t Distribution with 19 Degrees of Freedom

area = .005

−2.86   0   2.86 t-axis

Because −2.86 and +2.86 cut off areas of .005 in either tail of this distribution,

$$P\left\{-2.86 < \frac{\overline{X} - 61.8}{s/\sqrt{20}} < 2.86\right\} = .99$$

Or, equivalently,

$$P\left\{61.8 - 2.86\,(s/\sqrt{20}) < \overline{X} < 61.8 + 2.86\,(s/\sqrt{20})\right\} = .99$$

It follows that we should reject the null hypothesis at the $P = .01$ level of significance if either

$$\overline{x} \leq 61.8 - 2.86\,(s/\sqrt{20}) = \overline{X}_1^*$$

or

$$\overline{x} \geq 61.8 + 2.86\,(s/\sqrt{20}) = \overline{X}_2^*$$

But

$$\sum_{i=1}^{20} x_i^2 = 88878.12$$

so that

$$s = \sqrt{\frac{20(88878.12) - (1321.0)^2}{20(19)}}$$

$$= \sqrt{85.58} = 9.25$$

Therefore,

$$\bar{X}_1^* = 61.8 - 2.86\,(9.25/\sqrt{20}) = 55.9$$

$$\bar{X}_2^* = 61.8 + 2.86\,(9.25/\sqrt{20}) = 67.7$$

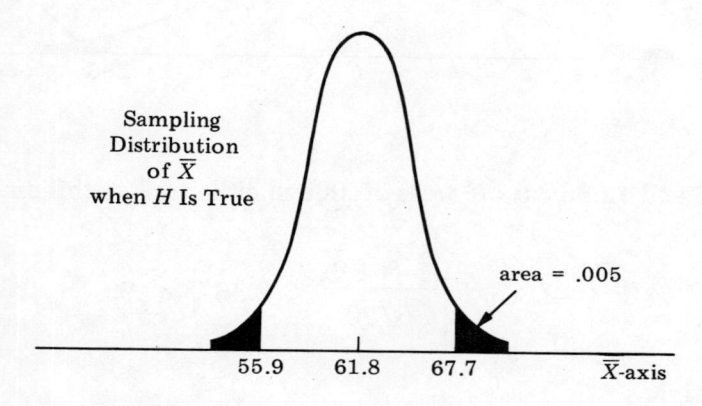

Sampling Distribution of $\bar{X}$ when H Is True

area = .005

55.9   61.8   67.7 $\bar{X}$-axis

Since $\bar{x}$ was 66.1—which falls between 55.9 and 67.7—our conclusion is to accept H.

Question 2.1.1

In the preceding analysis, the decision rule was expressed in terms of the sample mean; that is, we intended to reject H if $\bar{x}$ was either less than or equal to 55.9 or greater than or equal to 67.7. How would an equivalent decision rule be phrased in terms of the ratio

$$\frac{\bar{x} - 61.8}{9.25/\sqrt{20}} \;?$$

Question 2.1.2

For the data of Table 1, test

$$H: \mu = 61.8$$
vs.
$$A: \mu \neq 61.8$$

at the $P = .05$ level of significance.

Question 2.1.3

Test at the $P = .05$ level of significance whether the Shoshoni width-to-length ratios are compatible with Plato's standard.

Question 2.1.4

Suppose that the true standard deviation for Shoshoni ratios is known to be 9.0. If the true average ratio is 64.7, with what probability will a sample of size 20 fail to show that the Shoshonis have a different esthetic standard than the Greeks? Assume that H is being tested at the $P = .05$ level of significance.

Question 2.1.5

It can be shown mathematically that the value of w/l that satisfies the equation

$$\frac{w}{l} = \frac{l}{w + l}$$

is the limit of the continued fraction

$$\cfrac{1}{1 + \cfrac{1}{1 + \cfrac{1}{1 + \cfrac{1}{1 + \ldots}}}}$$

Evaluate the first six terms in this expression.

Question 2.1.6

Listed in Table 2 are 10 width-to-length ratios found on designs made by the Crow Indians. Test at the $P = .01$ level whether these are compatible with the golden rectangle.

Table 2
Width-to-Length Ratios ($\times$ 100)

58.7	58.8
85.0	86.2
52.2	44.7
66.6	57.1
52.9	48.3

Note:

$$\sum_{i=1}^{10} x_i = 610.50$$

$$\sum_{i=1}^{10} x_i^2 = 39{,}108.77$$

REFERENCE

1. DuBois, Cora, ed., *Lowie's Selected Papers in Anthropology* (University of California Press, 1960, pp. 137-142).

Example 2.2 Did she or didn't she?

Not long ago the following item appeared in Dear Abby's column:

> Dear Abby: You wrote in your column that a woman is pregnant for 266 days. Who said so? I carried my baby for ten months and five days, and there is no doubt about it because I know the exact date my baby was conceived. My husband is in the Navy and it couldn't have possibly been conceived any other time because I saw him only once for an hour, and I didn't see him again until the day before the baby was born.
>
> I don't drink or run around, and there is no way this baby isn't his, so please print a retraction about that 266-day carrying time because otherwise I am in a lot of trouble.
>
> *San Diego Reader*

Abby's answer was consoling and gracious but not very statistical:

> Dear Reader: The average gestation period is 266 days. Some babies come early. Others come late. Yours was late.

The question here is not whether the baby was late—that fact is already known. At issue is the credibility of the *length* of the delay. Ten months and five days is approximately 310 days, which means that the pregnancy exceeded the norm by 44 days. Is that figure compatible with the known variability of gestation times (in which case we accept the woman's claim), or is it so extreme a deviation that the only rational inference is that San Diego Reader is not telling the truth?

> **Objective**
>
> To establish the likelihood of a pregnancy being longer than ten months and five days.

Data

The observed pregnancy duration, X, was 310 days. Furthermore, medical records indicate that the true mean (μ) of the distribution of all pregnancy durations is 266 days; the true standard deviation (σ) is 16 days.

Analysis

This is not a typical hypothesis test for two reasons. First, our decision will have to be based on a single observation—namely, $x = 310$. Second, we already know the true value of μ—it equals 266. Nevertheless, by setting up the hypothesis as though we did *not* know μ and then carrying out the test in the usual way, we can establish the compatibility of μ with the observed x, and, in so doing, reach a decision as to whether we should believe or not believe San Diego Reader.

There is one other problem here. It will be necessary to assume that the distribution of pregnancy durations is normal. With a sample of size $n = 1$, the Central Limit Theorem cannot be used to approximate the distribution of $(\overline{X} - \mu)/(\sigma/\sqrt{n})$.

The hypotheses to be tested are

$$H: \quad \mu = 266$$

vs.

$$A: \quad \mu > 266$$

To reject H is to conclude that San Diego Reader is lying. Let $P = .01$ be the level of significance.

The null hypothesis will be rejected if fewer than 1% of all pregnancies exceed 310 days. There is a point, X^*, that represents the 99th percentile of the distribution of pregnancy durations.

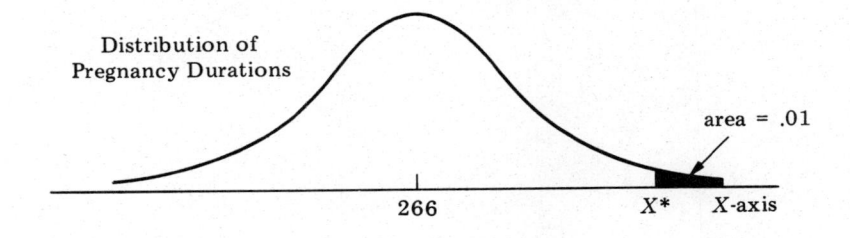

If 310 is greater than or equal to X^*, our conclusion will be to reject H.
To find X^*, note that

$$P(X > X^*) = .01$$

$$= P(\frac{X - 266}{16} > \frac{X^* - 266}{16}) = .01$$

$$= P(Z > \frac{X^* - 266}{16}) = .01$$

since the transformed variable $(X - \mu)/\sigma$ behaves like a standard normal variable. From tables of the standard normal, we have that

$$P(Z > 2.33) = .01$$

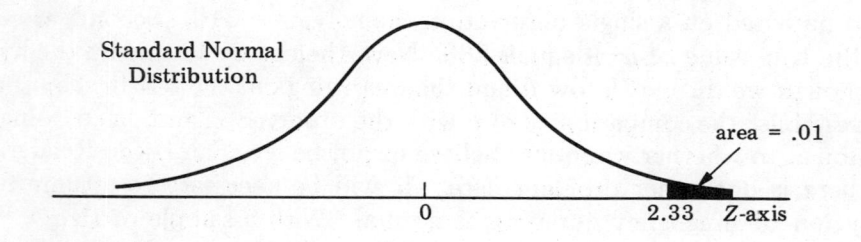

Standard Normal Distribution

area = .01

0   2.33   Z-axis

It must follow, then, from the last two equations that

$$\frac{X^* - 266}{16} = 2.33$$

Question 2.2.1

What conclusion is reached at the $P = .01$ level of significance?

Question 2.2.2

At what point in this analysis was it necessary to make the assumption that the distribution of pregnancy durations is normal?

Question 2.2.3

Test

$$H: \mu = 266$$
vs.
$$A: \mu > 266$$

at the $P = .005$ level of significance. At the $P = .001$ level of significance.

Question 2.2.4

Would it make sense to do this hypothesis test using a two-sided alternative? Explain.

Question 2.2.5

Compute the exact probability of a pregnancy exceeding 310 days (assuming that X is normal with $\mu = 266$ and $\sigma = 16$.)

Question 2.2.6

Mensa is an international society whose only requirement for membership is an IQ in the upper 2 per cent of the population. Assuming that IQ's are normally distributed with a mean of 100 and a standard deviation of 16, what is the lowest IQ that will qualify a person to belong to Mensa?

REFERENCE

1. *The Tennesseean*, Jan. 20, 1973.

Example 2.3 Polygraphs

Polygraphs are used in criminal investigations as a means of establishing the guilt or innocence of a suspect. They typically measure five functions: (1) thoracic respiration, (2) abdominal respiration, (3) blood pressure and pulse rate, (4) muscular movement and pressure, and (5) galvanic skin response. In principle, the magnitude of these responses when the subject is asked a relevant question ("Did you murder John Smith?") indicates whether he is lying or telling the truth.

Of course, the procedure is not infallible. The purpose of this study was to determine how accurate polygraphs really are under actual test conditions.

Objective

To determine the reliability of experts in interpreting the results of polygraph examinations.

Procedure

Seven experienced polygraph examiners were given a set of 40 records—20 were from innocent suspects and 20 from guilty suspects. The subjects had been asked 11 questions, on the basis of which each examiner was to make an overall judgment: "innocent" or "guilty."

Data

The seven polygraph examiners made a total of **280** decisions. The results are listed in the 2 × 2 table below.

		Suspect's True Status	
		Innocent	Guilty
Examiner's Decision	"Innocent"	131	15
	"Guilty"	9	125

Analysis

If the null hypothesis is taken to be the legal standard that a person is innocent until proven guilty, then each of the 280 decisions reduces to a choice between

H: suspect is innocent

vs.

A: suspect is guilty

As the table indicates, correct decisions were made 91.4 percent of the time

$$(= \frac{131 + 125}{280} \times 100).$$

Question 2.3.1

What percentage of the time were Type I errors committed? What percentage of the time were Type II errors committed?

Question 2.3.2

On the basis of these data, what level of significance would be associated with polygraph tests?

Question 2.3.3

In this particular setting, are Type I and Type II errors of equal consequence? Explain.

Question 2.3.4

Suppose a certain null hypothesis is being tested against a one-sided alternative. If H is rejected at the $P = .05$ level of significance, will it necessarily be rejected at the $P = .01$ level of significance? If H is rejected at the $P = .01$ level of significance, will it necessarily be rejected at the $P = .05$ level of significance?

Question 2.3.5

Which of the following statements is stronger?

1. H is rejected at the $P = .05$ level of significance.
2. H is rejected at the $P = .01$ level of significance.

REFERENCE

1. Horvath, Frank S., and John E. Reid, "The Reliability of Polygraph Examiner Diagnosis of Truth and Deception" (*Journal of Criminal Law, Criminology, and Police Science*, **62**, 1971, pp. 276-281).

Example 2.4 The Transylvania effect

In folklore, the full moon is often portrayed as something sinister, a kind of evil force that suppresses the Dr. Jekyll and brings out the Mr. Hyde in all of us. Over the centuries, many prominent writers and philosophers have shared this same belief. Milton, in *Paradise Lost*, refers to

> Demoniac frenzy, moping melancholy
> And moon-struck madness.

And Othello, after the murder of Desdemona, laments

> It is the very error of the moon,
> She comes more near the earth
>   than she was wont
> And makes men mad.

On a more scholarly level, Sir William Blackstone, the renowned eighteenth century English barrister, defined a lunatic as

> one who hath... lost the use of his reason and who hath lucid intervals, sometimes enjoying his senses and sometimes not, and that frequently depending upon the changes of the moon.

The possibility of lunar influences on human behavior is not without supporters among the scientific community. Studies done in recent years have attempted to link the "Transylvania effect," as it has come to be known, with higher suicide rates, pyromania, and even epilepsy. Often there have been hints of a correlation, but so far no definite conclusions have been reached.

In this example, we look at still another context in which the Transylvania effect might be expected to occur. The data are admission rates to the emergency room of a mental hospital.

Objective

To test whether admission rates to the emergency room of a mental hospital are higher during full moons than they are during the rest of the year.

Data

Table 1 shows the number of admissions to the emergency room of a Virginia mental health clinic during the 12 full moons from August 1971 to July 1972.

Table 1
Admissions During Full Moon

5.0	13.0
13.0	16.0
14.0	25.0
12.0	13.0
6.0	14.0
9.0	20.0

For the remainder of the year, the average number of daily admissions was 11.2.

Analysis

Let μ represent the true average number of admissions during the full moon. If there is *no* Transylvania effect, we would expect μ to equal 11.2, the average admission rate for the other days in the year. If there *is* a lunar effect, μ should be greater than 11.2. In the language of hypothesis testing, what we want to test is

$$H: \quad \mu = 11.2$$

vs.

$$A: \quad \mu > 11.2$$

If $x_1 = 5.0$, $x_2 = 13.0$, ..., $x_{12} = 20.0$, then

$$\sum_{i=1}^{12} x_i = 160.0 \qquad \sum_{i=1}^{12} x_i^2 = 2466.0$$

Therefore,

$$\bar{x} = \frac{160.0}{12}$$

$$= 13.3 \text{ admissions per full moon}$$

and

$$s = \sqrt{\frac{12(2466.0) - (160.0)^2}{12(11)}} = \sqrt{30.24}$$

$$= 5.5 \text{ admissions per full moon}$$

Suppose we decide to test H and A at the $P = .05$ level of significance. Because the behavior of

$$\frac{\bar{X} - 11.2}{s/\sqrt{12}}$$

is approximated by the Student t curve with 11 degrees of freedom (under the assumption that H is true), it follows that we should reject H if $(\bar{x} - 11.2)(s/\sqrt{12})$ is greater than or equal to 1.796.

Student t Distribution with 11 Degrees of Freedom

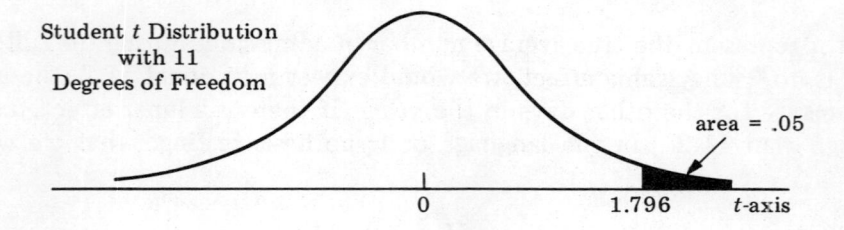

area = .05

0   1.796 t-axis

But

$$\frac{\bar{x} - 11.2}{s/\sqrt{12}} = \frac{13.3 - 11.2}{5.5/\sqrt{12}} = \frac{2.1}{1.59} = 1.32$$

which falls to the left of the critical value, so our conclusion is to accept the null hypothesis. Even though admissions during full moons were higher than they were for other days, the difference (= 13.3 - 11.2 = 2.1) was not statistically significant at the P = .05 level. Put another way, we have not established beyond reasonable doubt that the null hypothesis is false.

Question 2.4.1

Test these same hypotheses at the same level of significance using the procedure of Example 1.1. That is, find the numerical value of $\bar{X}^*$ and construct a decision rule accordingly.

Question 2.4.2

Test
$$H: \mu = 11.2$$
vs.
$$A: \mu > 11.2$$

at the $P = .10$ level of significance.

Question 2.4.3

Why was the alternative hypothesis chosen to be one sided in this example?

Question 2.4.4

What does it mean, specifically, to say that the behavior of

$$\frac{\overline{X} - 11.2}{s/\sqrt{12}}$$

is approximated by the Student t distribution with 11 degrees of freedom?

Question 2.4.5

If we had tested

$$H: \quad \mu = 11.2$$
vs.
$$A: \quad \mu \neq 11.2$$

at the $P = .05$ level of significance, under what conditions would we reject the null hypothesis?

Question 2.4.6

What other sorts of data might be collected to try to demonstrate the existence of the Transylvania effect?

REFERENCES

1. Blackman, Sheldon, and Don Catalina, "The Moon and the Emergency Room" (*Perceptual and Motor Skills*, 37, 1973, pp. 624-626).
2. Olvin, J. F., "Moonlight and Nervous Disorders" (*American Journal of Psychiatry*, **99**, 1943, pp. 578-584).

Example 2.5 Geochronometry

The chain of events that we call the geological evolution of the earth began hundreds of millions of years ago. Fossils have played a key role in documenting the *relative* times these events occurred, but to establish an *absolute* chronology scientists rely primarily on radioactive decay. In this example we look at the variability associated with a dating technique based on a mineral's potassium-argon ratio.

Almost all minerals contain potassium (K), as well as certain of its isotopes, including ^{40}K. But ^{40}K is unstable and decays into isotopes of argon and calcium, ^{40}A and ^{40}Ca. By knowing the rates at which these daughter products are formed and by measuring the amount of ^{40}A and ^{40}K present, it is possible to estimate the age of the mineral.

In the data we have seen thus far, location has been more important than dispersion in terms of the objective of the experiment. Recall, for instance, Example 2.1. There the null hypothesis was a statement that μ, the true average width-to-length ratios of rectangles preferred by the Shoshoni Indians was equal to the ratio preferred by the Greeks—namely, .618. Nothing explicit was said about the standard deviation of width-to-length ratios. Of course, it was necessary to know σ (or s) in order to reach a decision about μ, but still it was not a primary concern in its own right.

For characterizing an estimation technique, though, location and dispersion are *both* important. Clearly, if there are two competing techniques, the one with the smaller standard deviation would be preferred—all other factors being equal.

Objective

To analyze the variability associated with the potassium-argon method for dating minerals.

Data

Table 1 gives the estimated ages of 19 mineral samples collected from the Black Forest in southwestern Germany. Each of the estimates was made on the basis of the sample's potassium-argon ratio.

Analysis

Sometimes the amount of variability in a set of data is measured by dividing the standard deviation by the mean and multiplying the result by 100. This is known as the coefficient of variation and is abbreviated C.V.:

$$\text{C.V. (in per cent)} = \frac{s}{\bar{x}} \times 100$$

Table 1
Potassium-argon Dates
(Samples Taken from the Black Forest)

Specimen Number	Estimated Age (millions of years)
1	$x_1 = 249$
2	$x_2 = 254$
3	$x_3 = 243$
4	$x_4 = 268$
5	$x_5 = 253$
6	$x_6 = 269$
7	$x_7 = 287$
8	$x_8 = 241$
9	$x_9 = 273$
10	$x_{10} = 306$
11	$x_{11} = 303$
12	$x_{12} = 280$
13	$x_{13} = 260$
14	$x_{14} = 256$
15	$x_{15} = 278$
16	$x_{16} = 344$
17	$x_{17} = 304$
18	$x_{18} = 283$
19	$x_{19} = 310$

One of the earliest radioactive dating techniques—this one based on a mineral's lead content—is known to yield age estimates whose coefficient of variation is on the order of 11 per cent. We would like to know whether the variability characteristic of the potassium-argon method is significantly different.

The sample mean for the 19 ages given in Table 1 is

$$\bar{x} = \frac{\sum_{i=1}^{19} x_i}{19} = \frac{5{,}261}{19}$$

$$= 276.9 \text{ million years}$$

If the coefficient of variation for this sample were to be 11 per cent, the standard deviation would have to be

$$s = .11 \times 276.9$$

$$= 30.4 \text{ million years}$$

This suggests that we should test

$$H: \quad \sigma = 30.4$$

vs.

$$A: \quad \sigma \neq 30.4$$

where σ is the true standard deviation associated with potassium-argon dates. To accept the null hypothesis is to conclude that the potassium-argon method is comparable in precision to the method based on lead content.

Clearly, we should accept H only if the sample standard deviation, s, is sufficiently close to 30.4. Referring to Table 1, we find that

$$\sum_{i=1}^{19} x_i^2 = 1{,}469{,}945$$

which makes

$$s = \sqrt{\frac{19(1{,}469{,}945) - (5{,}261)^2}{19(18)}} = \sqrt{733.4}$$

$$= 27.1 \text{ million years}$$

At this point, the question to be resolved is whether the difference between σ and s—namely, $30.4 - 27.1 = 3.3$ million years—is small enough to be attributable to chance, in which case we should accept the null hypothesis, or large enough to *not* be, in which case we should reject the null hypothesis. Choosing between these two alternatives requires that we know the probabilistic behavior of s.

If H is true and if the population being sampled is more or less bell shaped, it can be shown mathematically that the behavior of

$$\frac{(n-1)s^2}{\sigma^2} = \frac{18 s^2}{\sigma^2}$$

is described by a χ^2 curve with $n - 1$ ($= 18$) degrees of freedom

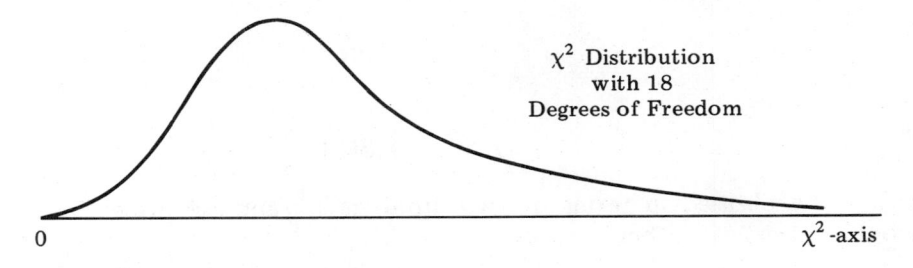

χ^2 Distribution with 18 Degrees of Freedom

χ^2-axis

Values of $(18s^2)/\sigma^2$ in either tail of the χ^2 distribution would constitute "proof" that H was false. For example, suppose we intended to test H versus A at the $P = .05$ level of significance. The numbers that cut off areas of .025 in either tail of the curve are 8.23 and 31.53.

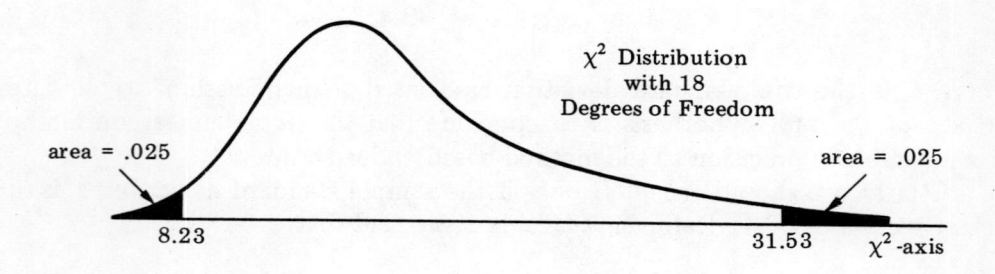

χ^2 Distribution with 18 Degrees of Freedom

Therefore, the null hypothesis should be rejected if $(18s^2)/\sigma^2$ is either

    (1) less than or equal to 8.23

or

    (2) greater than or equal to 31.53.

Question 2.5.1

At the $P = .05$ level of significance, what inference should we draw, accept H or reject H?

Question 2.5.2

Suppose the decision rule for testing

$$H: \quad \sigma = 30.4$$

vs.

$$A: \quad \sigma \neq 30.4$$

was to be phrased in terms of two numbers s_1^* and s_2^*. Specifically, we would reject H if either

$$(1) \quad s \leqslant s_1^*$$

or

$$(2) \quad s \geqslant s_2^*$$

where $P(s \leqslant s_1^*) = .025$ and $P(s \geqslant s_2^*) = .025$. What are the numerical values for s_1^* and s_2^*?

Question 2.5.3

Test

$$H: \quad \sigma = 30.4$$

vs.

$$A: \quad \sigma \neq 30.4$$

at the $P = .10$ level of significance.

Question 2.5.4

Test

$$H: \quad \sigma = 30.4$$

vs.

$$A: \quad \sigma < 30.4$$

at the $P = .10$ level of significance.

Question 2.5.5

In testing hypotheses about μ where the alternative is two sided, the critical values are equidistant from 0 (or from μ_0, depending on how the decision rule is phrased). However, the cutoff points for a two-sided test involving the standard deviation are *not* equidistant from the null hypothesis value for σ. Why?

REFERENCES

1. Carr, Donald P., and J. Laurence Kulp, "Potassium-Argon Method of Geochronometry" (*Bulletin of the Geological Society of America*, 68, 1957, pp. 763-784).
2. Lipson, Joseph, "Potassium-Argon Dating of Sedimentary Rocks" (*Bulletin of the Geological Society of America*, 69, 1958, pp. 137-150).
3. McIntyre, Donald B., "Precision and Resolution in Geochronometry," in *The Fabric of Geology*, Claude C. Albritton, Jr., ed. (Freeman, Cooper, and Co., p. 126).

Example 2.6 Echolocation

To hunt flying insects, bats emit high-frequency sounds and then listen for their echoes. Until an insect is located, these pulses are emitted at intervals of from 50 to 100 milliseconds. When an insect *is* detected, the pulse-to-pulse interval suddenly decreases—sometimes to as low as 10 milliseconds—thus enabling the bat to pinpoint its prey's position. This raises an interesting question: How far apart are the bat and the insect when the bat first senses that the insect is there? Or, in other words, what is the effective range of a bat's echolocation system?

This example is typical of the sort of situation where inferences take the form of confidence intervals rather than hypothesis tests. One of the primary objectives in this experiment would be to estimate μ, the true average bat-to-insect detection distance. However, there is no way to phrase this question in terms of a hypothesis test (as we did in Examples 2.1–2.5) because there is no "standard" value for μ that we can single out to associate with the null hypothesis. Presumably, this is the first time measurements such as these have ever been made.

Objective

To construct a confidence interval for the average distance separating a bat from a flying insect at the moment the bat's echolocation system first indicates the presence of the insect.

Procedure

In this experiment the technical problems involved in obtaining the data were far more difficult than the statistical problems involved in analyzing the data. The procedure that finally evolved was to put a bat into an 11 ft × 16 ft room, along with an ample supply of fruit flies and record the action with two synchronized 16mm sound-on-film cameras. By examining the two sets of pictures frame by frame, scientists could follow the bat's flight pattern and, at the same time, monitor its pulse frequency. For each insect that was eaten, it was possible to estimate the distance between the bat and the insect at the precise moment the bat's pulse-to-pulse interval decreased.

Data

Listed in Table 1 are the bat-to-insect detection distances recorded for 11 "catches."

Table 1

Catch Number	Detection Distance (cm)	Catch Number	Detection Distance (cm)
1	62	7	27
2	52	8	42
3	68	9	83
4	23	10	56
5	34	11	40
6	45		

Analysis

The parameter to be estimated by these data is μ, the overall average of the distribution of detection distances that would be recorded for all bats of this same species under these same conditions. The data in Table 1 are a random sample of size 11 taken from that distribution.

Letting $x_1 = 62$, $x_2 = 52$, ..., $x_{11} = 40$, we have that

$$\sum_{i=1}^{11} x_i = 532 \qquad \sum_{i=1}^{11} x_i^2 = 29{,}000$$

Therefore,

$$\bar{x} = \frac{532}{11}$$
$$= 48.4 \text{ cm}$$

and

$$s = \sqrt{\frac{11(29{,}000) - (532)^2}{11(10)}} = \sqrt{327.05}$$
$$= 18.1 \text{ cm}$$

Since the behavior of

$$\frac{\bar{X} - \mu}{s/\sqrt{11}}$$

is described by a Student t curve with 10 degrees of freedom,

$$P\left(-2.228 < \frac{\bar{X} - \mu}{s/\sqrt{11}} < 2.228\right) = .95$$

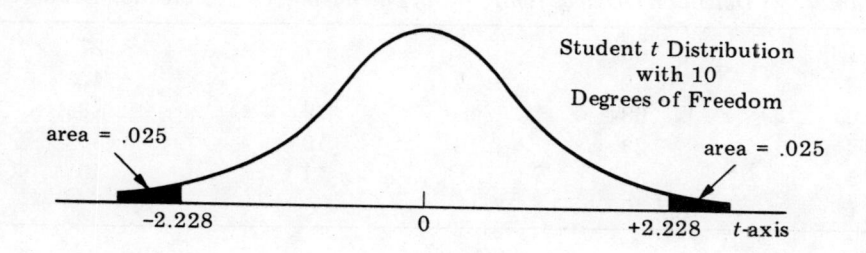

Student t Distribution with 10 Degrees of Freedom

area = .025   area = .025

−2.228   0   +2.228 t-axis

Equivalently,

$$P\left(\overline{X} - 2.228\frac{s}{\sqrt{11}} < \mu < \overline{X} + 2.228\frac{s}{\sqrt{11}}\right) = .95$$

implying that the range of values

$$\left(\overline{X} - 2.2228\frac{s}{\sqrt{11}}, \overline{X} + 2.228\frac{s}{\sqrt{11}}\right)$$

is a 95 per cent confidence interval for μ.

For these particular data, $\overline{x} = 48.4$ and $s = 18.1$. Therefore, the interval becomes

$$\left(48.4 - 2.228\frac{18.1}{\sqrt{11}}, 48.4 + 2.228\frac{18.1}{\sqrt{11}}\right)$$

$$= (48.4 - 12.2, 48.4 + 12.2$$

$$= (36.2 \text{ cm}, 60.6 \text{ cm})$$

It should be pointed out that the proper interpretation of a 95 per cent confidence interval is not what the name would imply. Specifically, it is *not* true that μ is contained in the interval (36.2 cm, 60.6 cm) 95 per cent of the time. Keep in mind that μ is a constant, so it lies in this particular interval either 100 per cent of the time or 0 per cent of the time. The 95 per cent refers to the confidence interval *procedure,* not to the particular outcome.

Suppose, for example, this entire experiment were repeated under similar conditions and a second set of 11 detection distances recorded. The mean and standard deviation for this second sample would almost surely be different than the values of $\overline{x}$ and s computed from Table 1. Consequently, a 95 per cent confidence interval based on these new data would have a lower limit different than 36.2 and an upper limit different than 60.6. Similarly, we can imagine taking a third sample, a fourth, and so on, each time computing a 95 per cent confidence interval for μ. The results could be summarized with a graph like the one shown in Figure 1.

Note that the true value of μ lies somewhere along the vertical axis. Its exact location, of course, always remains unknown. What we *do* know is that, in the long run, 95 per cent of all 95 per cent confidence intervals will contain μ as an interior point; the other 5 per cent will not. Unfortunately, we have no way of knowing whether *our* interval, (36.2 cm, 60.6 cm), is one of the 95 per cent or one of the 5 per cent.

Figure 1

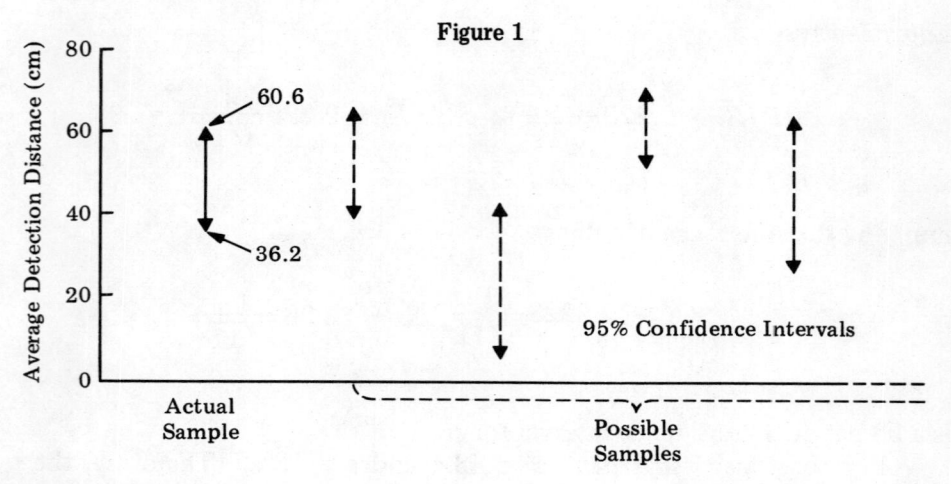

[Figure 1: Plot of Average Detection Distance (cm) vs. Actual Sample and Possible Samples, showing 95% Confidence Intervals. The actual sample interval ranges from 36.2 to 60.6 cm.]

Question 2.6.1

Construct 90 per cent and 99 per cent confidence intervals for these same data.

Question 2.6.2

For a given set of data, will a 99 per cent confidence interval be longer or shorter than a 95 per cent confidence interval? Is your answer to this question supported by your answer to Question 2.6.1?

Question 2.6.3

Suppose that two independent random samples of size n are taken from the same population. With the first sample we construct a 95 per cent confidence interval for μ; with the second sample, a 99 per cent confidence interval for μ. Is it possible that the 95 per cent interval will turn out to be longer than the 99 per cent interval? Explain.

Question 2.6.4

How would a *one-sided* 95 per cent confidence interval be constructed for the data of Table 1?

Question 2.6.5

Compute the coefficient of variation for the data of Table 1 (see Example 2.5). What does its magnitude imply about the precision of these data?

REFERENCE

1. Griffin, Donald R., Frederic A. Webster, and Charles R. Michael, "The Echolocation of Flying Insects by Bats" (*Animal Behavior*, 8, 1960, pp. 141-154).

Example 2.7 Scientific discoveries

Great discoveries in science tend to be made by persons who are quite young at the time. This is in marked contrast to the stereotype of the white-haired, absent-minded professor who may not know what day it is but somehow manages to know just about everything else. No doubt there have been such people, but the norm is in the other direction. Newton developed the calculus and formulated the theory of gravitation at the age of 23. Einstein showed that $E = mc^2$ and proposed the special theory of relativity when he was only 26.

In this example we look at 12 famous scientific discoveries and the ages of the scientists who made them.

Objective

To estimate the true average age at which great scientific discoveries are made.

Data

Listed in Table 1 are 12 major scientific breakthroughs from the middle of the sixteenth century to the early part of the twentieth century. All were revolutionary in terms of the impact they had on theories prevailing at the time.

Table 1

Discovery	*Discoverer*	*Date*	*Age*
Earth goes around sun	Copernicus	1543	40
Telescope, basic laws of astronomy	Galileo	1600	34
Principles of motion, gravitation, calculus	Newton	1665	23
Nature of electricity	Franklin	1746	40
Burning is uniting with oxygen	Lavoisier	1774	31
Earth evolved by gradual processes	Lyell	1830	33
Evidence for natural selection controlling evolution	Darwin	1858	49
Field equations for light	Maxwell	1864	33
Radioactivity	Curie	1896	34
Quantum theory	Planck	1901	43
Special theory of relativity, $E = mc^2$	Einstein	1905	26
Mathematical foundations for quantum theory	Schroedinger	1926	39

Question 2.7.1

Define the parameter to be estimated by these data.

Question 2.7.2

Let x_i denote the age of the scientist making the ith discovery, as listed in Table 1. What distribution describes the behavior of

$$\frac{\overline{X} - \mu}{s/\sqrt{12}} ?$$

Find the values of $-a$ and $+a$ such that

$$P\left(-a < \frac{\overline{X} - \mu}{s/\sqrt{12}} < a\right) = .95$$

Question 2.7.3

Write down the formula for a 95 per cent confidence interval for μ.

Question 2.7.4

Given that

$$\sum_{i=1}^{12} x_i = 425 \qquad \sum_{i=1}^{12} x_i^2 = 15{,}627$$

compute $\bar{x}$ and s.

Question 2.7.5

Substitute the values for $\bar{x}$ and s found in Question 2.7.4 into the formula for the 95 per cent confidence interval given in Question 2.7.3.

Question 2.7.6

Before constructing a confidence interval for a set of observations that extend over a long period of time—in this case, over almost 400 years—we should be convinced that there are no biases or trends in the x_i's. If, for example, there were a steady decrease from century to century in the age at which scientists made major discoveries, then the parameter μ would no longer be a constant and the confidence interval would be misleading. Draw a graph of the data in Table 1 by plotting "date of discovery" on the abscissa and "age of discoverer" on the ordinate. Does the variability in the x_i's appear to be random with respect to time?

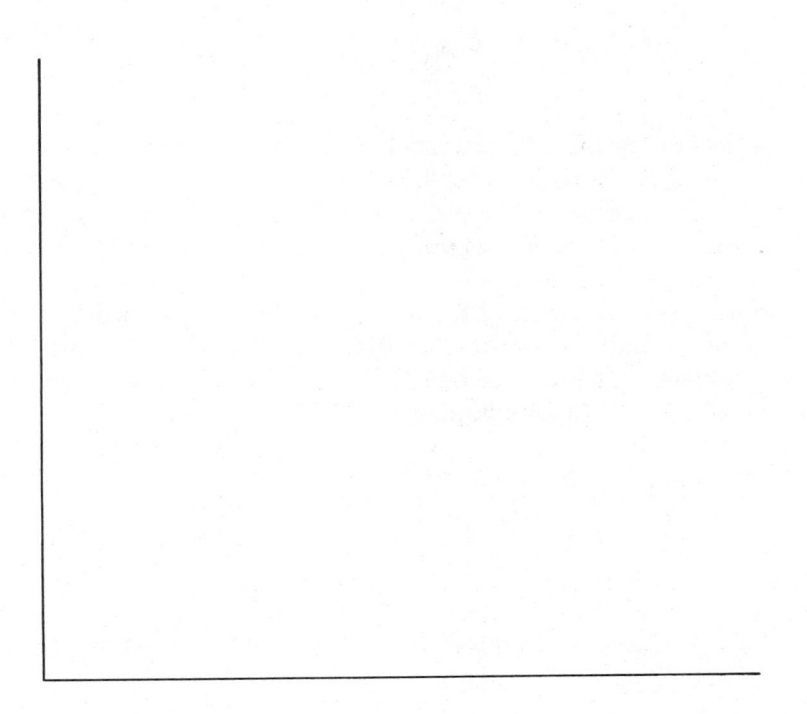

REFERENCE

1. Wood, Robert M., "Giant Discoveries of Future Science" (*Virginia Journal of Science*, 21, 1970, pp. 169–177).

2.3 Binomial data: the normal approximation (Examples 2.8–2.12)

Many experiments generate data that we would categorize as being binomial. A psychiatrist pronounces an accused murderer either "sane" or "insane"; a quality-control engineer inspects parts coming off an assembly line and grades each one on a "go" or "no-go" basis, depending on whether it meets certain specifications; a doctor examines a patient's electrocardiagram and judges it to be either "normal" or "abnormal"; a cloud-seeding experiment is either a "success" or a "failure" depending on whether it rains. Each of these events is a *binomial trial*: the measured response is nonnumerical and the number of possible responses is 2.

If a series of n binomial trials is observed, all under presumably the same conditions, we can record the total number of "successes" that occur. We will denote this number by the letter X. If the chances for a success are constant from trial to trial and are unaffected by (independent of) the outcomes of previous trials, we say that X is a *binomial random variable*. Furthermore, it can be shown that the probability distribution for X is given by the formula

$$P(X = x) = \binom{n}{x} p^x (1-p)^{n-x}, \quad x = 0, 1, \ldots, n$$

where p is the true probability of a success occurring at any given trial.

In statistical applications, n will be known and a certain value for X (namely, x) will have been observed. The problem will be to make some inference about p, either in the form of a hypothesis test or a confidence interval.

While inferences about p are occasionally made using the exact distribution of X as shown (see Example 2.8), n is usually so large that this is simply not feasible. The formula is too hard to evaluate. Instead, we need to rely on the Demoivre–Laplace statement of the Central Limit Theorem to approximate the distribution of X. All of this implies that if we were testing

$$H: \quad p = p_0$$

vs.

$$A: \quad p \neq p_0$$

where p_0 is some specified value for p, the decision rule would be based on the quantity

$$\frac{\frac{X}{n} - p_0}{\sqrt{\frac{p_0(1-p_0)}{n}}}$$

If the null hypothesis is true, this has approximately the standard normal distribution, and critical values are gotten accordingly. Applications of this result are detailed in Examples 2.9 and 2.10.

To construct confidence intervals for p, we use the fact that the behavior of

$$\frac{\frac{X}{n} - p}{\sqrt{\frac{(X/n)(1-(X/n))}{n}}}$$

is also approximated by the standard normal curve. Since $P(Z \leqslant -1.96) = .025$ and $P(Z \geqslant 1.96) = .025$

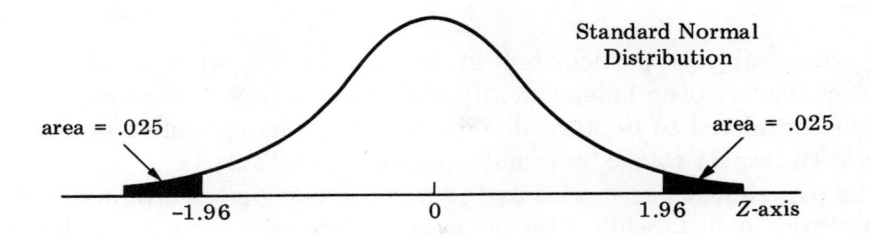

we can say that

$$P\left(-1.96 < \frac{\frac{X}{n} - p}{\sqrt{\frac{(X/n)(1-(X/n))}{n}}} < 1.96\right) = .95$$

Therefore,

$$\left((x/n) - 1.96\sqrt{\frac{(x/n)(1-(x/n))}{n}},\ (x/n) + 1.96\sqrt{\frac{(x/n)(1-(x/n))}{n}}\right)$$

is a 95 per cent confidence interval for p.

Example 2.8 Handwriting analysis

To a handwriting analyst (or graphologist), the way you cross your t's or put loops on your p's and q's, or form any letter reveals a lot about the way you think and act. Presumably, all these handwriting characteristics when interpreted properly can tell whether a person is emotional, egotistical, sarcastic, self-reliant, generous—even whether he is a credit risk.

Naturally, claims like these are difficult to verify (how do you prove, for example, that someone is self-reliant?), and not everyone believes that graphologists can do what they say they can. This example describes a small-scale experiment where a well-known handwriting expert was put to the test.

Objective

To test whether a professional graphologist can distinguish a normal person's handwriting from that of a psychotic.

Procedure

Twenty subjects participated in the study; ten were considered to be psychotics (having been independently diagnosed as such by two physicians) and ten were considered to be normal. The normals were chosen to match the psychotics with respect to age, sex, and education (see Table 1).

The experiment began with each psychotic being given five minutes to read a short story. A little while later he was asked to write down as much of it as he could remember. Then the experimenter read the normal member of each psychotic-normal pair what the psychotic had written, and he (the normal) transcribed it in his own handwriting. When this was done, the two handwriting samples for each of the pairs were put into a folder, in random order.

The graphologist was then called in and presented with the ten folders. Her objective was to pick out the sample in each of the folders that had been written by the psychotic.

Table 1
Subjects

Folder Number	Diagnosis	Sex	Age	Education
1	Schizophrenia	M	28	High School
	Normal	M	31	High School
2	Schizophrenia	M	20	High School
	Normal	M	21	High School
3	Schizophrenia	M	32	College, 1 yr
	Normal	M	27	College, 2 yrs
4	Schizophrenia	M	43	High School
	Normal	M	47	High School
5	Schizophrenia	M	25	College
	Normal	M	30	College
6	Schizophrenia	F	26	College, 1 yr
	Normal	F	26	College, 2 yrs
7	Manic Depressive	M	63	College
	Normal	M	58	College
8	Schizophrenia	M	23	High School
	Normal	M	21	High School
9	Schizophrenia	F	39	Grammar School
	Normal	F	39	Grammar School
10	Dementia Praecox	F	62	Grammar School
	Normal	F	65	Grammar School

Data

The graphologist made correct matchups in 6 of the 10 folders.

Analysis

The question to be answered here is whether this particular graphologist has any special ability for distinguishing the handwriting of normal persons from that of psychotics. Suppose we let p denote her probability of doing so. That is,

$p = P$ (any particular psychotic-normal pair will be correctly identified)

If she has *no* ability, p will be $\frac{1}{2}$; if she does, p will be greater than $\frac{1}{2}$. It follows, then, that the hypotheses to be tested are

$$H: p = \frac{1}{2}$$

vs.

$$A: p > \frac{1}{2}$$

Ultimately, the choice between H and A will be made on the basis of X, the number of correct identifications made in the ten trials. Note that from the way the experiment has been set up, we can expect X to behave like a binomial random variable (why?). Therefore,

$$P(X = r) = \binom{10}{r} p^r (1-p)^{10-r}, \text{ for } r = 0, 1, \ldots, 10$$

Furthermore, if we assume that H is true

$$P(X = r) = \binom{10}{r} \left(\frac{1}{2}\right)^r \left(1 - \frac{1}{2}\right)^{10-r}$$

$$= \binom{10}{r} \left(\frac{1}{2}\right)^{10}, \text{ for } r = 0, 1, \ldots, 10$$

To test

$$H: p = \frac{1}{2}$$

vs.

$$A: p > \frac{1}{2}$$

at, say, the $P = .05$ level of significance, we need to find the value X^* such that

$$P(X \geq X^*) \leq .05$$

and

$$P(X \geq X^* - 1) > .05$$

when the null hypothesis is true.

Question 2.8.1

Find the value of X^* referred to previously and carry out the hypothesis test. What would be the value of X^* if we were to test H versus A at the $P = .01$ level of significance? At the $P = .10$ level of significance?

Question 2.8.2

How likely is it that a person would make six or more correct identifications just by chance?

Question 2.8.3

What is the probability of the graphologist getting all ten correct, assuming she is only guessing? Suppose the 20 samples were all mixed together. What probability would she then have of correctly identifying the ten psychotics?

Question 2.8.4

As a side experiment, a group of 25 persons untrained in either psychology or graphology were presented with the same folders and given the same instructions. Their "scores" are listed in Table 2.

Table 2
Handwriting Identifications

Number of Correct Guesses	Number of People
2	2
3	1
4	4
5	7
6	8
7	2
8	1
	25

Compute the sample mean for these data. Assuming the responses of these 25 people can be described by a binomial distribution, estimate the value of p.

Question 2.8.5

Using the answer to Question 2.8.4, compute the "expected" numbers (= 25 $P(X = r)$) of the 25 persons making r correct guesses ($r = 0, 1, 2, \ldots, 10$).

Table 3

Number Correct, x	$P(X = x)$	25 $P(X = x)$
0		
1		
2		
3		
4		
5		
6		
7		
8		
9		
10		

Question 2.8.6

Draw histograms on the same graph for the observed and expected frequencies given in Tables 2 and 3. Would you conclude that the binomial distribution adequately describes the abilities of these 25 persons to analyze handwriting?

Question 2.8.7

Test

$$H: p = \frac{1}{2}$$

vs.

$$A: p > \frac{1}{2}$$

for the data of Table 2.

REFERENCE

1. Pascal, Gerald, and Barbara Suttell; "Testing the Claims of a Graphologist" (*Journal of Personality*, **16**, 1947, pp. 192-197).

Example 2.9 Death months

There is a theory that people tend to "postpone" their deaths until after some event having particular significance to them has passed. Occasions such as a birthday, a family holiday, or the return of a loved one have all been suggested as the sorts of events that might have this effect. Politics might be another. A recent study has shown that the mortality rate in the United States drops considerably during the Septembers and Octobers of presidential election years.

All of this may sound rather far-fetched but the fact is that the postponement theory is supported by evidence that is difficult to explain away. Some of that evidence is discussed in this example.

As in Example 2.8, the data in this study are binomial. But there is a major difference between the two. In the handwriting example the number of binomial trials was quite small (10), which meant that the entire probability distribution for X could be easily computed. But here the number of trials is large (348), and the only feasible way of doing a hypothesis test is to make use of the Central Limit Theorem.

Objective

To see whether a person's death month is influenced by his birth month.

Data

Table 1 lists the birth months and death months of 348 persons, all listed in *Four Hundred Notable Americans*. (The other 52 were persons who either had not died yet or whose biographical information was incomplete.)

Analysis

If the postponement theory is correct, there should be a tendency for people *not* to die in the month immediately preceding their birth month. We see from Table 1 that of the 38 people who were born in January, one died in December; of the 32 born in February, one died in January. Adding these two deaths to the other 10 starred entries in Table 1, we find that a total of 16 persons (or $(16/348) \times 100 = 4.6$ per cent) died in the month preceding their birth month.

If it is assumed that the chances of dying in any one month are equal, then $\frac{1}{12}$ (= 8.3 per cent) of all deaths would be expected to occur, by chance, in the months prior to the 348 birth months. How can we explain, then, the deficit that

Table 1
Birth Months and Death Months of 348 American Celebrities

Born	Jan	Feb	Mar	Apr	May	Jun	Jul	Aug	Sep	Oct	Nov	Dec
Jan	1	2	5	7	4	4	4	4	2	4	0	1*
Feb	1*	3	6	6	4	0	0	4	2	2	2	2
Mar	2	1*	5	3	2	4	3	4	1	2	0	2
Apr	1	3	3*	2	2	5	4	4	0	3	2	1
May	2	1	1	1*	1	1	3	2	2	2	1	2
Jun	2	0	0	3	2*	1	3	2	0	2	1	1
Jul	4	2	5	3	4	1*	4	3	2	2	0	4
Aug	3	1	1	1	1	2	1*	3	4	3	3	1
Sep	1	2	2	3	3	1	6	1*	2	3	3	4
Oct	4	2	5	2	2	2	4	1	0*	1	3	0
Nov	2	6	3	4	1	4	2	2	5	4*	1	2
Dec	4	4	1	4	5	0	5	0	2	5	0*	2

actually occurred (an observed 4.6 per cent as compared to an expected 8.3 per cent)? There are two possibilities:

1. The deviation of the observed death percentage (4.6 per cent) from 8.3 per cent is *small enough* to be accounted for by sampling variability.
2. The deviation of 4.6 per cent from 8.3 per cent is *large enough* to constitute substantial proof that something other than chance—specifically, the postponement theory—was operable.

The decision to believe either (1) or (2) will be made by calculating the likelihood of observing, by chance, a percentage as small as, or smaller than, 4.6 even though the true death percentage was 8.3. In the framework of hypothesis testing, the choices would be expressed as

$$H: p = .083$$

vs.

$$A: p < .083$$

where p is the actual proportion of persons dying in the month preceding their birth month. If X denotes the *number* of persons (out of 348) dying "a month early"—and if H is assumed to be true—the sampling distribution of $X/348$ will be approximated by a normal curve with a mean equal to .083 and a standard deviation equal to $\sqrt{(.083)(.917)/348} = .015$ (Why?). Therefore, at, say, the $P = .01$ level of significance, H should be rejected if $x/348$ is more than 2.33 standard deviations of $X/348$ *to the left* of .083. That is, we reject H if $x/348$ is less than or equal to $(X/348)^*$, where

$$(X/348)^* = .083 - 2.33\sqrt{(.083)(.917)/348}$$

$$= .083 - .034$$

$$= .049$$

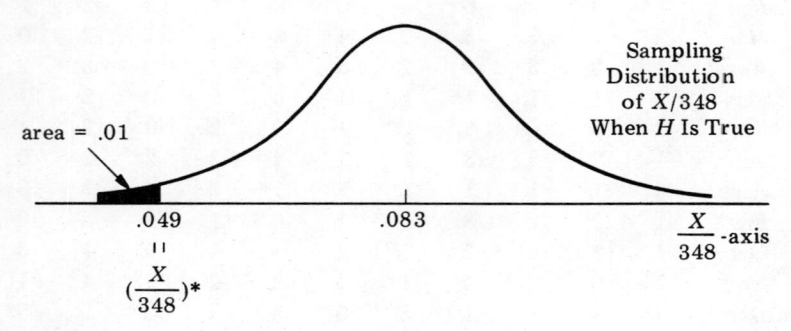

[Figure: Sampling Distribution of $X/348$ When H Is True, with area = .01 shaded to the left of .049; center at .083; $X/348$-axis; $(X/348)^* = .049$]

Question 2.9.1

At the $P = .01$ level, what conclusion do we reach? What conclusion would we reach at the $P = .005$ level?

Question 2.9.2

The following listing of 1,202 birth months and death months was taken from Hickok's *Who Was Who in American Sports*. Is there any indication in these data that the postponement theory is correct?

Table 2
Birth Months and Death Months of 1,202 American Athletes

	Died											
Born	*Jan*	*Feb*	*Mar*	*Apr*	*May*	*Jun*	*Jul*	*Aug*	*Sep*	*Oct*	*Nov*	*Dec*
Jan	11	1	12	4	6	14	6	8	4	6	8	11
Feb	10	10	10	8	8	7	8	9	6	13	7	8
Mar	12	9	8	13	8	4	8	7	10	5	17	7
Apr	14	5	6	12	11	12	8	8	10	11	4	7
May	11	6	16	3	9	9	5	2	12	6	3	11
Jun	11	5	6	7	9	10	1	9	4	8	5	4
Jul	14	10	6	9	6	3	10	6	7	8	5	14
Aug	9	12	8	9	7	5	11	8	5	10	17	10
Sep	10	14	6	8	13	6	2	8	13	10	6	7
Oct	6	18	10	15	5	10	17	6	8	7	8	2
Nov	13	9	5	7	10	7	6	10	8	7	8	8
Dec	9	5	12	10	10	4	9	6	11	8	7	6

Question 2.9.3

Do the data of Table 2 support the hypothesis that people tend to postpone their deaths until after New Year's Day?

Question 2.9.4

The assumption that a person is equally likely to die in any of the 12 months is not entirely true. In 1966, for example, deaths in the U.S. followed the month-by-month pattern shown in Table 3. (In other years, the pattern was similar.) How might this information be incorporated into the analysis called for in Question 2.9.3?

Table 3
U.S. Deaths (1966)

Month	Number
Jan	166,761
Feb	151,296
Mar	164,804
Apr	158,973
May	156,455
Jun	149,251
Jul	159,924
Aug	145,184
Sep	141,164
Oct	154,777
Nov	150,678
Dec	163,882

Question 2.9.5

Another way to determine whether deaths occur at random with respect to births is to compare the *day* of birth and the *day* of death for persons who were born and who died in the same month. For the 107* people falling into this category, a total of 50 died "after" their birthday.

Let p be the true proportion of persons whose death day comes after birthday. If chance is the only factor that needs to be considered, p should equal $\frac{1}{2}$. Test

$$H: p = \frac{1}{2}$$

vs.

$$A: p \neq \frac{1}{2}$$

at the $P = .05$ level of significance.

---

*Only 107 persons are included here rather than the 112 who appear along the diagonal of Table 2 (11 in Jan-Jan + 10 in Feb-Feb, and so forth) because four people were born and died on the same day of the month and the exact days for a fifth were not known.

REFERENCES

1. Hickok, Ralph, *Who Was Who in American Sports* (Hawthorn Books, Inc., 1971).
2. Phillips, David P., "Deathday and Birthday: An Unexpected Connection," in *Statistics: A Guide to the Unknown*, Judith M. Tanur, et al., ed. (Holden-Day, 1972, pp. 52-65).
3. *The Tennesseean*, Nov. 3, 1974, p. 24.

Example 2.10   Honeybees and paper flowers

Nature has provided flowers that are pollinated by bees with various biological mechanisms for insuring that the job gets done properly. Many, for example, have strongly scented pollen; others are brightly colored. What may be another device, one certainly not so obvious, is the set of radiating lines sometimes found on petals. It has been suggested that these rays serve as "nectar guides," leading bees from the outside of the flower to the inside where the pollen is stored. The following experiment was designed to test that theory.

Objective

To see whether bees have an affinity for flowers whose petals have lines.

Procedure

First, a number of honeybees were trained to feed off a special table. The top of this table consisted of two panes of glass; pressed between them were a number of irregularly-shaped paper flowers, approximately 3 cm in diameter (see

Figure 1a). Drops of sugar syrup had been placed on the table top above each of the flowers. A swarm of bees were then released near the table and allowed to feed off as many "flowers" as they desired.

After the bees had become accustomed to this setup, the syrup was removed and the original flowers were replaced with a grid of 16 new flowers, eight *with* lines and eight *without* (see Figure 1b). When the bees were reintroduced, a count was made of the number landing on each of the patterns. The intention was to use the proportion of times that bees landed on flowers *with* lines as a means of testing whether the nectar guide theory had any merit.

Figure 1

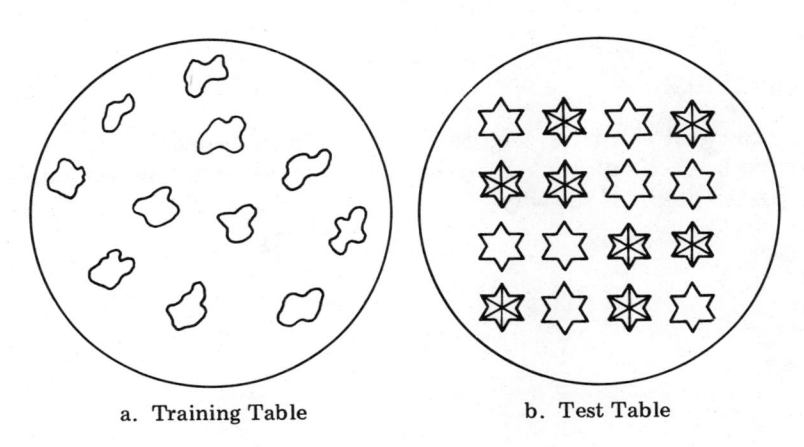

a. Training Table   b. Test Table

Data

A total of 107 bees landed on the test table—66 on flowers with lines and 41 on flowers without lines.

Question 2.10.1

Formulate and carry out an appropriate hypothesis test for this problem.

Question 2.10.2

In the grid shown in Figure 1, each pattern appeared twice in each row and twice in each column. Do you think this was the deliberate intention of the experimenter? If so, why?

Question 2.10.3

If more than one bee was allowed to land on the second table at the same time, would the binomial model still apply? Explain.

Question 2.10.4

How would you test the hypothesis that the bees prefer the periphery of the table to the interior?

REFERENCE

1. Free, J. B., "Effect of Flower Shapes and Nectar Guides on the Behavior of Foraging Honeybees" (*Behavior*, 37, 1970, pp. 269-285).

Example 2.11 A Gallup poll

Survey sampling is an area of statistics that profoundly affects our everyday lives. Market researchers solicit consumer opinions, which eventually become translated into what we can and cannot buy in the supermarket; politicians speak out on public issues only after reading how their constituency responded to the latest Harris survey; even what we watch on television is determined by an ongoing audience survey.

One of the most common statistical problems associated with any survey is the estimation of the true proportion of the population that would give a certain response to a particular question. Knowing that 45% of a sample of 200 persons favor Proposition A, what are some reasonable values for the proportion of the entire population that support it? Is it likely, for example, that the true proportion might be as high as 50%? Or as low as 35%? Often, as in this example, the answer to such a question takes the form of a confidence interval.

Objective

To construct a confidence interval for the binomial parameter, p.

Statistics in the Real World

Procedure

In August of 1972, a Gallup poll was conducted to determine current attitudes toward birth control education in the public high schools. The respondents were 1,345 men and women of voting age; all were white. The question was

> Would you approve or disapprove of having nationwide programs of birth control education in public high schools?

Data

A total of 955 (= 71 per cent) said they would approve.

Analysis

If it can be assumed that the opinions of these 1,345 persons constitute a random sample (see Question 2.11.5), we can construct a confidence interval for p, the nationwide approval rate. Recall that if X denotes the number of persons out of 1,345 approving the proposal, the sampling distribution of

$$\frac{\frac{X}{1,345} - p}{\sqrt{\frac{(\frac{X}{1,345})(1 - \frac{X}{1,345})}{1,345}}}$$

will be approximately normal. Therefore,

$$Pr\left\{-1.96 < \frac{\frac{X}{1,345} - p}{\sqrt{\frac{(\frac{X}{1,345})(1 - \frac{X}{1,345})}{1,345}}} < 1.96\right\} = .95$$

which implies that a 95 per cent confidence interval for p will have the form

$$\left(\frac{X}{1,345} - 1.96\sqrt{\frac{(\frac{X}{1,345})(1 - \frac{X}{1,345})}{1,345}}, \frac{X}{1,345} + 1.96\sqrt{\frac{(\frac{X}{1,345})(1 - \frac{X}{1,345})}{1,345}}\right)$$

The particular interval that would be gotten from *this* sample, since X was 955, is

$$\left(\frac{955}{1{,}345} - 1.96\sqrt{\frac{\frac{955}{1{,}345}(1-\frac{955}{1{,}345})}{1{,}345}},\ \frac{955}{1{,}345} + 1.96\sqrt{\frac{\frac{955}{1{,}345}(1-\frac{955}{1{,}345})}{1{,}345}}\right)$$

$= (.71 - 1.96\sqrt{.000153},\ .71 + 1.96\sqrt{.000153}$

$= (.71 - .02,\ .71 + .02)$

$= (.69, .73)$

Question 2.11.1

What is the precise interpretation that we associate with intervals such as (.69, .73)? What information is gained by estimating the unknown proportion in favor with an interval as opposed to a single point?

Question 2.11.2

Construct a 50 per cent confidence interval for these same data. Construct a 99 per cent confidence interval.

Question 2.11.3

Write down a general expression for the *length* of a 95 per cent confidence interval for p. For fixed n, when will the interval be longest?

Question 2.11.4

What is the smallest sample size required to guarantee that a 99 per cent confidence interval for p will be no longer than .07?

Question 2.11.5

Support for birth control education programs is not uniform over all sectors of society. "Age" is a particularly critical factor; the percentage of young people approving of such programs is much higher than the corresponding percentage for older people. Knowing that to be true, does it make sense to construct a confidence interval for the entire population? If so, how should the sample be chosen?

REFERENCE

1. Blake, Judith, "The Teenage Birth Control Dilemma and Public Opinion" (*Science*, 180, 1973, pp. 708-712).

Example 2.12 The spiral aftereffect

The spiral aftereffect is an optical illusion that may be observed by staring at something that exhibits fairly rapid and continuous motion, like a waterfall or traffic on a thruway—and then looking away; for an instant, everything will appear to move in the opposite direction.

In laboratory settings, the motion needed to create the spiral aftereffect can be provided by an Archimedes spiral.

When rotated, the arms of the spiral will appear to be moving *towards* the center. At the instant the spiral is stopped, anyone perceiving the effect will think the lines, momentarily, reversed direction and began moving *away* from the center.

When describing a phenomenon quantitatively, it helps to be able to demonstrate that the response distribution approximates a well-known probability model. Sometimes this can be accomplished by showing that the conditions surrounding the response variable are very similar to the assumptions specified in the model. Another way is to show, after the fact, that the histogram of the response distribution has very nearly the same shape, location, and dispersion as the histogram for a particular probability model. Both ways play a part in the following example.

Objective

To see whether the abilities of people to perceive the spiral aftereffect can be described by a binomial distribution.

Procedure

A sample of 420 children, ranging in age from 5 to 16, were the subjects. Each was instructed to stare at a rotating Archimedes spiral for 30 seconds and then asked to comment on what happened when the motion stopped. If the subject correctly perceived the illusion, he was given a score of 1; if he didn't, his score was 0. This procedure was repeated eight times, so the final scores could range from 0 to 8.

Question 2.12.1

If the final score for each person was to be treated as a binomial random variable, what would be the numerical value for n? What would be the range of values for X? What meaning would be attached to the parameter p? Write down a general formula for $P(X = r)$.

Data

Figure 2 shows the observed distribution of scores.

Figure 2

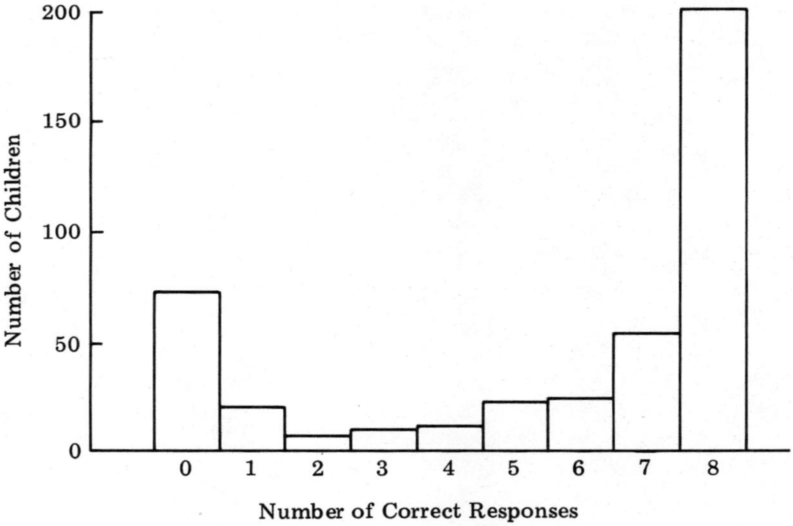
200
150
100
50
0
Number of Children
0
1
2
3
4
5
6
7
8
Number of Correct Responses

Question 2.12.2

Estimate the parameter p with the data of Figure 2.

Question 2.12.3

Compute the probability distribution of X using the values of n and p found in Questions 2.12.1 and 2.12.2. Superimpose a histogram of the "expected" X distribution over the histogram of the observed X distribution in Figure 2.

Expected X Distribution

r	$420\,P(X = r)$

Question 2.12.4

How would you account for the differences between the two histograms of Figure 2? Is the binomial distribution a suitable model for describing spiral after-effect perception?

REFERENCES

1. Blau, Theodore H., and Robert E. Schaffer, "The Spiral Aftereffect Test (SAET) As a Predictor of Normal and Abnormal Electroencephalographic Records in Children" (*Journal of Consulting Psychologists*, 24, 1960, pp. 35-42).
2. Holland, Harry C., *The Spiral After-Effect* (Pergamon Press, 165, pp. 1-26).

Chance fights ever on the side of
the prudent.
Euripides

Chapter 3

The Two-sample Problem

3.1 Introduction

"Does Macy's tell Gimbel's?" was a famous slogan of years past that summed up the spirit of competition that prevailed between those two New York stores. It implied, among other things, that it was not so much what Macy's did that mattered, but, rather, what Macy's did *relative* to what Gimbel's did. This same desire to compare is the reason why two-sample problems arise in so many experimental situations. Ralph Nader notwithstanding, the management at Ford is likely to be more concerned with how their cars perform relative to Chevys and less about whether they meet any absolute standards. This may not be a good attitude to take but it seems to be a statistical fact of life.

In a typical two-sample problem, there are two independent samples, $(x_1, \ldots, x_n)$ and $(y_1, y_2, \ldots, y_m)$, representing the two theoretical population distribution P_X and P_Y. For example, P_X and P_Y might be the distributions of gas mileages that Fords and Chevys could theoretically get under certain conditions, with $(x_1, x_2, \ldots, x_n)$ and $(y_1, y_2, \ldots, y_m)$ being the actual mileages recorded for n Fords and m Chevys. Or P_X and P_Y might be the theoretical distributions of spatial-perception scores that would be made by males and females, respectively, on a standard IQ test.

But regardless of the specifics, the basic question to be answered in these problems is whether P_X and P_Y are the same. Admittedly, they can be different in a variety of ways—they can have different shapes, or different locations, or different dispersions. For most situations, though, shifts in location have the most physical relevance, and we usually find ourselves testing H: $\mu_X = \mu_Y$, where μ_X and μ_Y are the means of P_X and P_Y, respectively.

One of the examples in this section reexamines the Twain-Snodgrass data of Chapter 1. There we tried to resolve the authorship question by visually comparing word-length frequency polygons based on writings from the two (?) authors. Here we take a more quantitative approach—but end up drawing the same conclusion. Another example in this chapter describes a chemical analysis done on twelfth century Byzantine coins and concludes that silver shortages were a problem even then. A third example describes a very unusual bumper sticker survey.

Examples 3.1 through 3.5 are all illustrations of the *two-sample t test*. In Example 3.6, we focus on a different set of parameters, σ_X and σ_Y, the standard deviations of P_X and P_Y. This also qualifies as a two-sample problem but its analysis is quite different. The test statistic is s_X^2/s_Y^2 and the critical values come from an F distribution. The last two examples show how to compare two sets of binomial data.

3.2 The two-sample t test (Examples 3.1–3.5)

The basic statistical tool for the two-sample problem is the two-sample t test. Given data that consist of two independent random samples, $(x_1, x_2, \ldots, x_n)$ and $(y_1, y_2, \ldots, y_m)$, drawn from population distributions P_X and P_Y with means and standard deviations equal to μ_X, μ_Y, σ_X, and σ_Y, the two-sample t test is a method for testing

$$H: \mu_X = \mu_Y$$

vs.

$$A: \mu_X \neq \mu_Y$$

(or A: $\mu_X < \mu_Y$ or A: $\mu_X > \mu_Y$ if the alternative is one sided). In its usual form it assumes that (1) P_X and P_Y are both approximately bell shaped and (2) σ_X and σ_Y are equal, but unknown. When these conditions are satisfied and when the null hypothesis is true, it can be shown that the sampling distribution of

$$\frac{\overline{X} - \overline{Y}}{s_p \sqrt{(1/n) + (1/m)}}$$

(where s_p = pooled standard deviation =

$$\sqrt{\frac{(n-1)s_X^2 + (m-1)s_Y^2}{n+m-2}}\,)$$

is described by a Student t curve with $n + m - 2$ degrees of freedom. This implies that we should accept or reject H: $\mu_X = \mu_Y$ according to the magnitude of the difference between the sample means, $\bar{x} - \bar{y}$.

In a typical, real-life situation, n and m will not be large enough to allow us to test whether P_X and P_Y are, in fact, bell shaped but we *can* examine the second assumption, that $\sigma_X = \sigma_Y$. We can test

$$H: \quad \sigma_X = \sigma_Y$$

vs.

$$A: \quad \sigma_X \neq \sigma_Y$$

by making use of the fact that when H is true the ratio of the sample variances, s_X^2/s_Y^2, has an F distribution with $n = 1$ and $m - 1$ degrees of freedom. To be strictly correct, a test of H: $\sigma_X = \sigma_Y$ should precede every test of H: $\mu_X = \mu_Y$. If it happens that H: $\sigma_X = \sigma_Y$ is rejected, we can still test H: $\mu_X = \mu_Y$—but only after the t statistic is slightly modified. Occasionally, a test of H: $\sigma_X = \sigma_Y$ will be warranted even when μ_X and μ_Y are of no interest. Example 3.6 describes one such instance.

Example 3.1 Statistics and the coin collector

In 1965, a silver shortage in the United States prompted Congress to authorize the minting of silverless dimes and quarters. They also recommended that the silver content of half dollars be reduced from 90 per cent to 40 per cent. Historically, fluctuations in the amount of rare metals found in coins are not uncommon. The data in this example compare the silver contents of a certain twelfth century Byzantine coin minted at two different times during the reign of Manuel I (1143-1180).

Statistical studies such as this can add an entirely new dimension to archeological investigations that would otherwise be purely descriptive. Oftentimes, chemical analyzes of coin alloys or pottery compositions can provide scientists with important clues in reconstructing a society's political and cultural history.

Objective

To test whether the average silver content of a Byzantine coin was the same for two different mintings.

Procedure

The type of coin examined in this study was first struck in 1092 as part of a monetary reformation instituted by Alexius I. The analysis consisted of first dissolving chips from a particular specimen in a solution that was 50 per cent nitric acid. Then this solution was titrated with sodium chloride until all the silver chloride had precipitated out. By weighing the precipitate it was possible to calculate the percentage of silver that was in the coin.

Data

The specimens analyzed in this study were part of a large hoard discovered not too long ago in Cyprus. Nine coins were identified as being part of the first coinage made during the reign of Manuel I. The other seven came from a fourth coinage, minted several years later.

Table 1
Silver Content of Coins
Minted During Reign of Manuel I (in %)

First coinage	Fourth coinage
5.9	5.3
6.8	5.6
6.4	5.5
7.0	5.1
6.6	6.2
7.7	5.8
7.2	5.8
6.9	
6.2	

Analysis

Presumably, the nine coins in Table 1 associated with the first coinage are, in fact, a random sample of *all* coins minted at that particular time. Likewise, the seven specimens in the second sample accurately represent the population of coins minted during the fourth coinage. Suppose we let μ_X and μ_Y represent the true average silver contents characteristic of the first and fourth coinages, respectively. The hypotheses that we want to test are

$$H:\ \mu_X = \mu_Y$$

vs.

$$A:\ \mu_X \neq \mu_Y$$

Clearly, the null hypothesis should be rejected if $\bar{x} - \bar{y}$, the difference between the two sample means, is either much larger than 0 or much smaller than 0. But in order to know precisely what "much larger" and "much smaller" mean in numerical terms, we need to relate the amount of variability *between* the samples (as measured by $\bar{x} - \bar{y}$) to the amount of variability *within* the samples. This latter variability is measured by a quantity known as the pooled standard deviation,

The Two-sample Problem

s_p. As the following formula shows, s_p is the square root of the weighted average of the individual sample variances:

$$s_p = \sqrt{\frac{(n-1)s_X^2 + (n-1)s_Y^2}{n+m-2}}$$

Here n and m are the two sample sizes.

If $H: \mu_X = \mu_Y$ is true, the probabilistic behavior of the ratio

$$\frac{\overline{X} - \overline{Y}}{s_p \sqrt{\frac{1}{n} + \frac{1}{m}}}$$

is approximated by the Student t distribution with $n + m - 2$ degrees of freedom. In this case, $n + m - 2 = 9 + 7 - 2 = 14$.

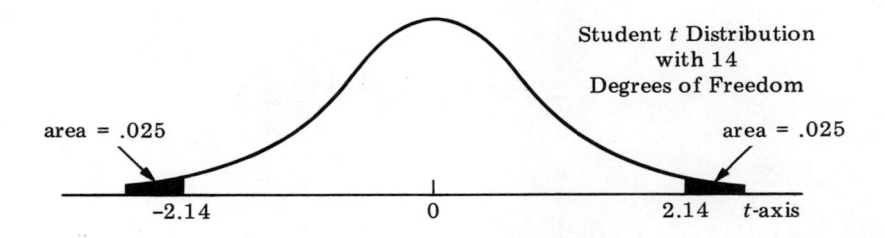

Student t Distribution with 14 Degrees of Freedom

area = .025   area = .025

−2.14   0   2.14 t-axis

If H and A were to be tested at the $P = .05$ level of significance, H would be rejected if

$$(\overline{x} - \overline{y})/(s_p \sqrt{\tfrac{1}{9} + \tfrac{1}{7}})$$

was either

    1. less than or equal to −2.14

or

    2. greater than or equal to +2.14.

From Table 1,

$$\sum_{i=1}^{9} x_i = 60.7 \qquad \sum_{i=1}^{9} x_i^2 = 411.75$$

so that

$$\bar{x} = \frac{60.7}{9} = 6.7$$

and

$$s_X^2 = \frac{9(411.75) - (60.7)^2}{9(8)} = .30$$

Also,

$$\sum_{i=1}^{7} y_i = 39.3 \qquad \sum_{i=1}^{7} y_i^2 = 221.43$$

making

$$\bar{y} = \frac{39.3}{7} = 5.6$$

and

$$s_Y^2 = \frac{7(221.43) - (39.3)^2}{7(6)} = .13$$

Therefore, the pooled standard deviation is given by

$$s_p = \sqrt{\frac{8(.30) + 6(.13)}{9 + 7 - 2}} = \sqrt{.227} = .48$$

Substituting these values for $\bar{x}$, $\bar{y}$, s_p, n, and m into the test statistic gives

$$\frac{6.7 - 5.6}{.48\sqrt{\frac{1}{9} + \frac{1}{7}}} = \frac{1.1}{.48\sqrt{.254}} = \frac{1.1}{.24}$$

$$= 4.58$$

Since 4.58 lies to the right of the upper critical values (2.14), our decision is to reject the null hypothesis. The average amount of silver used in coins made during the fourth minting was different (specifically, it was *less*) than the average amount used in the first minting.

Question 3.1.1

Express the $P = .05$ decision rule in terms of $\bar{x} - \bar{y}$. That is, what ranges of $\bar{x} - \bar{y}$ values result in the null hypothesis being rejected at the $P = .05$ level of significance?

Question 3.1.2

What would the decision rule be for testing

$$H: \mu_X = \mu_Y$$

vs.

$$A: \mu_X > \mu_Y$$

at the $P = .05$ level of significance?

Question 3.1.3

Why would it not be reasonable always to define the pooled standard deviation as the square root of the unweighted average of the sample variances:

$$s_p = \sqrt{\frac{s_X^2 + s_Y^2}{2}} \ ?$$

Question 3.1.4

Construct a 95 per cent confidence interval for the difference between the population means, $\mu_X - \mu_Y$. Hint: use the fact that

$$\frac{\overline{X} - \overline{Y} - (\mu_X - \mu_Y)}{s_p \sqrt{\frac{1}{n} + \frac{1}{m}}}$$

has a Student t distribution with $n + m - 2$ degrees of freedom.

Question 3.1.5

Four coins known to have been minted during the third coinage in the reign of Manuel I were found to have silver contents equal to 4.9 per cent, 5.5 per cent, 4.6 per cent, and 4.5 per cent. The average silver content for these four coins is 4.9 per cent, which is different than the 5.6 per cent found for the sample from the fourth coinage (see Table 1). Can we conclude that the true average silver contents characteristic of the third and fourth coinages were different?

REFERENCE

1. Hendy, M. F., and J. A. Charles, "The Production Techniques, Silver Content and Circulation History of the Twelfth-Century Byzantine Trachy" (*Archaeometry*, 12, 1970, pp. 13–21).

Example 3.2 Quintus Curtius Snodgrass revisited

In Example 1.1 we tried to decide whether ten letters published in the *New Orleans Daily Crescent* in 1861 were written by their acknowledged author, Quintus Curtius Snodgrass, or by their suspected author, Mark Twain. The analysis was simple. A visual comparison was made between three word-length distributions constructed from writings known to be the work of Mark Twain and one constructed from the letters in question. As it turned out, there appeared to be more agreement *within* the Twain samples than between the Twain samples and the Snodgrass sample. On that basis, it was concluded, very tentatively, that Twain did not write the Snodgrass letters.

Except in extreme cases—and this was not an extreme case—a subjective analysis of this sort will always leave some doubters. Can we be sure that the observed differences in word-length patterns are not within normal limits of sampling variability? How similar would the frequency polygons have to be for us to reach the opposite conclusion—that the authors were the same? What is clearly needed is a more quantitative approach, one that can assess in a very precise way the probability of all the writings being the work of a single author.

Objective

To assess, quantitatively, the likelihood of Mark Twain being the author of the Quintus Curtius Snodgrass letters.

Data

One way to get a more objective solution to this problem is to focus on just one piece of the word-length frequency distributions—say, "three-letter words"—and test whether the proportion of three-letter words used by Twain is equal to the corresponding proportion for Snodgrass. Table 1 shows the three-letter word proportions found in the eight Twain letters making up Sample 1, as referred to in Example 1.1; also listed are the corresponding figures for the ten Snodgrass letters.

Table 1
Proportion of Three-letter Words

Twain	Proportion	QCS	Proportion
Sergeant Fathom Letter	.225	Letter I	.209
Madame Caprell Letter	.262	Letter II	.205
Mark Twain Letters in		Letter III	.196
Territorial Enterprise		Letter IV	.210
First Letter	.217	Letter V	.202
Second Letter	.240	Letter VI	.207
Third Letter	.230	Letter VII	.224
Fourth Letter	.229	Letter VIII	.223
First *Innocents Abroad* Letter		Letter IX	.220
First Half	.235	Letter X	.201
Second Half	.217		

Analysis

If $x_1 = .225$, $x_2 = .2621, \ldots, x_8 = .217$ and $y_1 = .209$, $y_2 = .205, \ldots, y_{10} = .201$, then

$$\bar{x} = \frac{1.855}{8} = .232$$

and

$$\bar{y} = \frac{2.097}{10} = .210$$

To analyze these data, we need to decide what the magnitude of the difference between the sample means, $\bar{x} - \bar{y} = .232 - .210 = .022$, actually tells us. Specifically, if μ_X and μ_Y denote the true proportions of three-letter words written by Twain and Snodgrass, respectively, does an observed sample difference of .022 imply that the two are *not* the same? Or is .022 small enough to still be compatible with the hypothesis that they are? More formally, we must choose between

$$H: \mu_X = \mu_Y$$

vs.

$$A: \mu_X \neq \mu_Y$$

Since

$$\sum_{i=1}^{8} x_i^2 = .4316$$

and

$$\sum_{i=1}^{10} y_i^2 = .4406$$

the two sample variances are

$$s_X^2 = \frac{8(.4316) - (1.855)^2}{8(7)}$$

$$= .0002103$$

and

$$s_Y^2 = \frac{10(.4406) - (2.097)^2}{10(9)}$$

$$= .0000955$$

It follows that the pooled standard deviation is given by

$$s_p = \sqrt{\frac{7(.0002103) + 9(.0000955)}{8 + 10 - 2}}$$

$$= \sqrt{.0001457}$$

$$= .012$$

Recall that if $H: \mu_X = \mu_Y$ is true, the sampling distribution of

$$\frac{\overline{X} - \overline{Y}}{s_p \sqrt{(1/8) + (1/10)}}$$

will be approximated by a Student t curve with 16 degrees of freedom. Therefore,

$$P\left\{-2.92 < \frac{\overline{X} - \overline{Y}}{s_p \sqrt{(1/8) + (1/10)}} < 2.92\right\} = .99 \quad \text{(Why?)}$$

This means that for a test at the $P = .01$ level of significance, H should be rejected if $\overline{x} - \overline{y}$ is either less than or equal to $(\overline{X} - \overline{Y})_1^*$ or greater than or equal to $(\overline{X} - \overline{Y})_2^*$, where

$$(\overline{X} - \overline{Y})_1^* = -2.92\, s_p \sqrt{(1/8) + (1/10)}$$

$$= -2.92(.012)\sqrt{.225}$$

$$= -.017$$

and

$$(\overline{X} - \overline{Y})_2^* = 2.92\, s_p \sqrt{(1/8) + (1/10)}$$

$$= .017$$

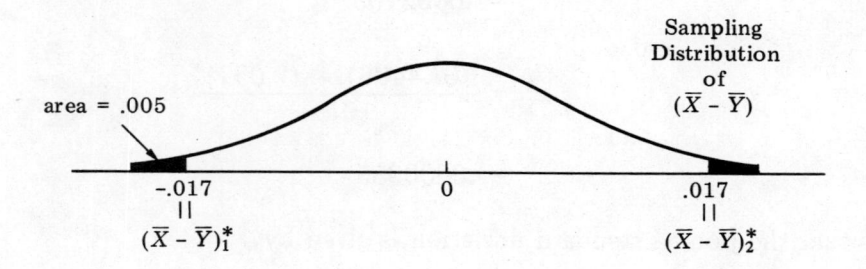

Question 3.2.1

What conclusion do we reach at the $P = .01$ level of significance? At the $P = .001$ level of significance?

Question 3.2.2

Phrase the $P = .01$ decision rule for this problem in terms of

$$\frac{\overline{X} - \overline{Y}}{s_p \sqrt{\frac{1}{8} + \frac{1}{10}}}$$

Question 3.2.3

Table 2 gives the proportions of four-letter words for the same samples indicated in Table 1.

Table 2
Proportion of Four-letter Words

Twain	Proportion	QCS	Proportion
Sergeant Fathom Letter	.187	Letter I	.176
Madame Caprell Letter	.212	Letter II	.170
Mark Twain Letters in		Letter III	.163
Territorial Enterprise		Letter IV	.159
First Letter	.196	Letter V	.168
Second Letter	.179	Letter VI	.172
Third Letter	.189	Letter VII	.181
Fourth Letter	.208	Letter VIII	.167
First *Innocents Abroad* Letter		Letter IX	.181
First Half	.195	Letter X	.220
Second Half	.185		

Test

$$H: \mu_X = \mu_Y$$

vs.

$$A: \mu_X \neq \mu_Y$$

Use the $P = .01$ level of significance. What do μ_X and μ_Y refer to here? Note that

$$\sum_{i=1}^{8} x_i = 1.551 \qquad \sum_{i=1}^{8} x_i^2 = .3016$$

$$\sum_{i=1}^{10} y_i = 1.757 \qquad \sum_{i=1}^{10} y_i^2 = .3113$$

Question 3.2.4

Would it be reasonable to use a one-sided alternative in Question 3.2.1? Explain.

Question 3.2.5

Construct a 99 per cent confidence interval for $\mu_X - \mu_Y$, using the data of Table 2. What relationship exists between a $100(1-P)$ per cent confidence interval for $\mu_X - \mu_Y$ and a test of

$$H: \mu_X = \mu_Y$$
vs.
$$A: \mu_X \neq \mu_Y$$

at the P level of significance?

REFERENCE

1. Brinegar, Claude S., "Mark Twain and the Quintus Curtius Snodgrass Letters: A Statistical Test of Authorship" (*Journal of the American Statistics Association*, 58, 1963, pp. 85-96).

Example 3.3 Hospital carpeting

For the most part, the decisions we have reached in this chapter and in the previous chapter have not involved any major economic consequences. Questions arising over disputed authorship, the Transylvania effect, the percentage of silver in Byzantine coins, and so on are largely academic in nature. Oftentimes though, statistical methods (in particular, hypothisis tests) are brought to bear on matters much more practical.

Consider, for example, the following question, which continues to be an issue faced by health administrators all over the country: Should the floors in a hospital be carpeted? There is no doubt that the installation of carpeting has certain esthetic merits. But what about the health consequences? Is a vacuumed carpet as sanitary as a washed floor? A strong "no" would carry a lot of weight in the administrator's final decision.

Objective

To compare the levels of airborne bacteria in carpeted and uncarpeted patient rooms.

Procedure

One way to assess the sanitary consequences of carpeting in a hospital environment is to compare bacteria counts taken directly from the carpeting with those taken directly from tile. Another way, one perhaps more relevant to the spread of infection, is to compare levels of airborne bacteria in carpeted and uncarpeted rooms. This can be done by pumping room air at a known rate over a growth medium, incubating that medium, and then counting the number of bacteria colonies that form. In this study, done in a Montana hospital, room air was forced over two Petri dishes at the rate of one cubic foot per minute.

Data

Sixteen patient rooms were tested, eight were carpeted and eight were uncarpeted. All were similar with respect to the amount of traffic each one had and the kinds of housekeeping services each one was provided. Table 1 and Figure 1 show the results.

Table 1
Levels of Airborne Bacteria

Carpeted rooms	Number of colonies per/ft^3 of air	Uncarpeted rooms	Number of colonies per/ft^3 of air
212	$x_1 = 11.8$	210	$y_1 = 12.1$
216	$x_2 = 8.2$	214	$y_2 = 8.3$
220	$x_3 = 7.1$	215	$y_3 = 3.8$
223	$x_4 = 13.0$	217	$y_4 = 7.2$
225	$x_5 = 10.8$	221	$y_5 = 12.0$
226	$x_6 = 10.1$	222	$y_6 = 11.1$
227	$x_7 = 14.6$	224	$y_7 = 10.1$
228	$x_8 = 14.0$	229	$y_8 = 13.7$

Figure 1
Airborne Bacteria in Patient Rooms

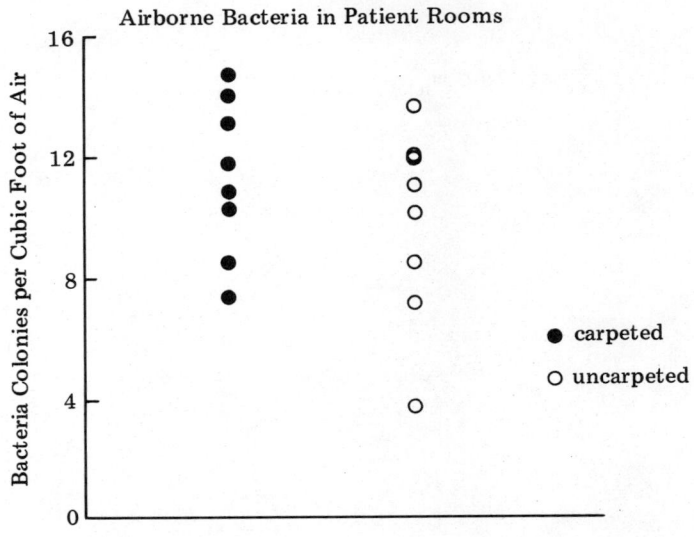

Question 3.3.1

Define the parameters to be tested and state the null and alternative hypotheses.

Question 3.3.2

Given that

$$\sum_{i=1}^{8} x_i = 89.6 \qquad \sum_{i=1}^{8} x_i^2 = 1{,}053.70$$

and

$$\sum_{i=1}^{8} y_i = 78.3 \qquad \sum_{i=1}^{8} y_i^2 = 838.49$$

compute $\bar{x}$, $\bar{y}$, s_X^2, s_Y^2, and s_p.

Question 3.3.3

Sketch a diagram of the Student t curve with $8 + 8 - 2 = 14$ degrees of freedom. Indicate which values of

$$\frac{\overline{X} - \overline{Y}}{s_p \sqrt{(1/8) + (1/8)}}$$

would result in the null hypothesis being rejected. Assume that .05 is the level of significance.

Question 3.3.4

Use the answers to Question 3.3.2 to evaluate the decision rule formulated in Question 3.3.3.

Question 3.3.5

Test the hypotheses of Question 3.3.1 at the $P = .10$ level of significance.

Question 3.3.6

For a given set of data, is it possible for $H: \mu = \mu_0$ to be rejected in favor of $A: \mu \neq \mu_0$ at the $P = .0001$ level of significance while $H: \mu = \mu_0$ is accepted in favor of $A: \mu < \mu_0$ at the $P = .10$ level of significance? Explain.

REFERENCE

1. Walter, William G., and Angie Stobie, "Microbial Air Sampling in a Carpeted Hospital" (*Journal of Environmental Health*, 30, 1968, p. 405).

Example 3.4 The Thematic Apperception Test

One of the standard personality inventories used by psychologists is the Thematic Apperception Test (TAT). It consists of a series of pictures depicting a number of everyday experiences. The subject is asked to examine the pictures carefully and to make up a story about each one. Interpreted properly, the content of these stories can provide valuable insights into the subject's mental well-being.

In order for a test such as this to have any credibility, it must be able to distinguish one personality type from another. This means that the distributions of TAT scores achieved by persons having, say, Personality Type A and Personality Type B must differ with respect to either shape, location, or dispersion (or some combination of all three). In the example described here, this idea is carried one step further. We are given two groups of children representing two distinct personality types, and the question is whether or not their *mothers* will score differently on the Thematic Apperception Test.

Objective

To see whether mothers of schizophrenic children respond differently to the Thematic Apperception Test than mothers of normal children.

Procedure

A total of 40 mothers participated, 20 of whom had schizophrenic children. Each mother was shown the same set of 10 pictures, and the stories they made up were categorized according to the sort of parent-child relationship each one exhibited. The data in Table 1 give the number of those stories (out of 10) showing a *positive* parent-child relationship, one where the mother was clearly capable of interacting with her child in a flexible, open-minded way.

Table 1
Number of Stories Showing a Positive Parent-child Relationship (out of 10)

	Normal Children				Schizophrenic Children			
Mother	Score	Mother	Score	Mother	Score	Mother	Score
1	$x_1 = 8$	11	$x_{11} = 2$	21	$y_1 = 2$	31	$y_{11} = 0$
2	$x_2 = 4$	12	$x_{12} = 1$	22	$y_2 = 1$	32	$y_{12} = 2$
3	$x_3 = 6$	13	$x_{13} = 1$	23	$y_3 = 1$	33	$y_{13} = 4$
4	$x_4 = 3$	14	$x_{14} = 4$	24	$y_4 = 3$	34	$y_{14} = 2$
5	$x_5 = 1$	15	$x_{15} = 3$	25	$y_5 = 2$	35	$y_{15} = 3$
6	$x_6 = 4$	16	$x_{16} = 3$	26	$y_6 = 7$	36	$y_{16} = 3$
7	$x_7 = 4$	17	$x_{17} = 2$	27	$y_7 = 2$	37	$y_{17} = 0$
8	$x_8 = 6$	18	$x_{18} = 6$	28	$y_8 = 1$	38	$y_{18} = 1$
9	$x_9 = 4$	19	$x_{19} = 3$	29	$y_9 = 3$	39	$y_{19} = 2$
10	$x_{10} = 2$	20	$x_{20} = 4$	30	$y_{10} = 1$	40	$y_{20} = 2$

Analysis

This example illustrates the similarity between Student t curves and the standard normal. We have already seen that there is, in fact, an entire *family* of Student t curves, each indexed by a parameter we call "degrees of freedom." As this parameter increases, the corresponding t curves become more and more similar to the standard normal.

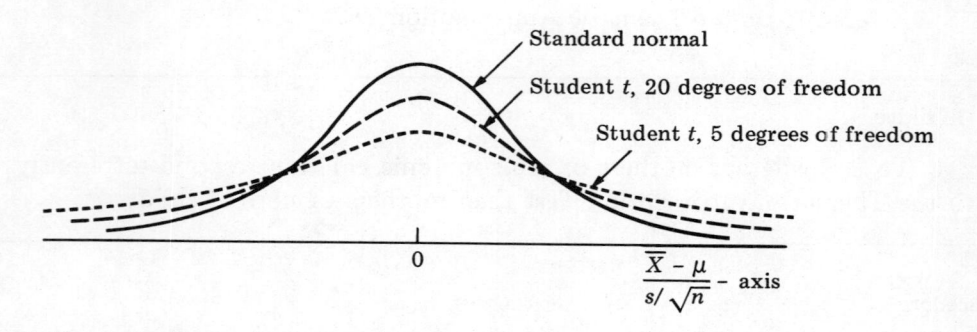

When the degrees of freedom get as high as 30, the curves are the same for all practical purposes. For the data of Table 1, $n + m - 2 = 20 + 20 - 2 = 38$. Therefore, the behavior of

$$\frac{\overline{X} - \overline{Y}}{s_p \sqrt{\frac{1}{20} + \frac{1}{20}}}$$

can be satisfactorily approximated by a standard normal.

Question 3.4.1

Graph the data of Table 1.

Question 3.4.2

Set up the appropriate null and alternative hypotheses for this problem. Hint: Is it likely, in general, that the mother of a schizophrenic child will have a *better* relationship with her child than the mother of a normal child?

Question 3.4.3

Test at the $P = .05$ level of significance the hypotheses formulated in Question 3.4.2. Phrase the decision rule in two different ways.

Question 3.4.4

Can it be concluded from this study that the attitudes of mothers of schizophrenic children are the *cause* of their parent-child relationships being less favorable?

Question 3.4.5

Suppose that $\mu_X - \mu_Y$ were really 1.2. Sketch a diagram that shows the probability of making a Type II error if the decision rule of Question 3.4.3 is used.

REFERENCE

1. Werner, Martha, James R. Stabenau, and William Pollin, "Thematic Apperception Test Method for the Differentiation of Families of Schizophrenics, Delinquents, and 'Normals'" (*Journal of Abnormal Psychology*, 75, 1970, pp. 139-145).

Example 3.5 Walking exercises for the newborn

Baby books are always filled with "first" dates—when the baby first crawled, or first stood up, or said his first word. Surprisingly, our knowledge of these developmental stages is quite limited. We know their norms, but very little else. For the most part, medical researchers have simply not pursued the various factors and conditions that may influence a child's early growth. The following study is an exception.

Here we look at one of the most important of these firsts: when a baby first walks alone. Normally, this does not occur until the child is almost 14 months old. The theory suggested in this example is that it might be possible to condition infants so that they would attain this particular developmental level at a significantly earlier age.

Objective

To determine whether special "walking" exercises in the newborn can lower the average age at which infants first walk alone.

Procedure

Twelve one-week-old male infants from middle and upper-middle class families were used as subjects. All were white. The birth order of the infants and the ages of the parents were kept as similar as possible to guard against possible biases.

Two groups of size six were randomly chosen. The first, or active-exercise group, received four three-minute sessions of special walking and placing exercises each day, beginning with the second week and lasting through the eighth week. Infants in the second, or passive-exercise group, received the same overall amount of daily exercise but were not given the walking and placing training. This second group acted as a control (see Question 3.5.3). In each instance, the exercises were conducted in the child's home by his parents. After eight weeks the program was discontinued.

Data

Table 1 shows the ages, in months, when the children first walked by themselves.

Table 1
First Walking Times (months)

Active-exercise Group (x)	Passive-exercise Group (y)
9.00	11.00
9.50	10.00
9.75	10.00
10.00	11.75
13.00	10.50
9.50	15.00

Question 3.5.1

Graph the data of Table 1.

Question 3.5.2

Formulate and test an appropriate hypothesis for these data. Define all parameters. Note that

$$\sum_{i=1}^{6} x_i = 60.75 \qquad \sum_{i=1}^{6} x_i^2 = 625.56$$

$$\sum_{i=1}^{6} y_i = 68.25 \qquad \sum_{i=1}^{6} y_i^2 = 794.31$$

Question 3.5.3

Notice that five of the six infants in the passive-exercise group eventually walked at a much earlier age than the 14-month mean. What might account for this? What does this imply about using the one-sample model to test

$$H: \quad \mu = 14$$

vs.

$$A: \quad \mu < 14$$

as a means of establishing the effectiveness of the special walking and placing exercises? What would μ represent in this case?

Question 3.5.4

It was mentioned in the procedure section that the 12 subjects were divided randomly into two groups of size six. This is usually done using a random number table. First the subjects are numbered—in this case, from 01 to 12. Then we scan down two consecutive columns in a random number table until six numbers between 01 and 12 are located. The subjects having these six numbers become the members of the first sample. Suppose we were given the following set of random numbers. Which infants would be put into the active-exercise group?

$$62015$$
$$07729$$
$$15435$$
$$11239$$
$$87106$$

$$10276$$
$$04940$$
$$33385$$
$$12992$$
$$02745$$

REFERENCE

1. Zelazo, Philip R., Nancy Ann Zelazo, and Sarah Kolb, " 'Walking' in the Newborn" (*Science*, **176**, pp. 314-315).

Example 3.6 Glacial flow

In Chapter 2 we saw that hypothesis tests for one-sample problems generally involve a location parameter, either μ or p. Sometimes, though, as in the case of Example 2.5, it was more reasonable to focus on the standard deviation. These same priorities hold for two-sample problems. Almost always the null hypothesis will be $H: \mu_X = \mu_Y$ (or $H: p_X = p_Y$ if the data are binomial). But occasionally a situation arises where a test of $H: \sigma_X = \sigma_Y$ versus $A: \sigma_X \neq \sigma_Y$ would be more relevant. The study described in this example is one such instance.

The easiest way to measure the movement, or flow, of a glacier is with a camera. First a set of reference points are established at various sites near the edge of the glacier. Then these points, and the glacier, are photographed from an airplane. The problem is this: How long should the time interval be between photographs? If too *short* a period has elapsed, the glacier will not have moved very far, and the errors associated with the photographic technique would be relatively large. If too *long* a period has elapsed, parts of the glacier might be deformed by the surrounding terrain. This could introduce substantial variability in the velocity estimates from point to point.

In this example two sets of flow rates are examined, one based on photographs taken three years apart, the other, five years apart. Both sets of data were made on the same glacier and under the same conditions.

Objective

To compare the variabilities in estimated glacial flow rates based on photographs taken three years apart as opposed to five years apart.

Procedure

The target for this study was the Hoseason Glacier in the Antarctic. A total of seven points along the glacier's edge was photographed in August 1957, and again in October 1960. A second set of five points was taken in October 1960, and in January 1965. From other considerations it could be assumed that the flow of the glacier was constant for the eight years in question.

Data

By comparing "before" and "after" pictures, it was possible to estimate the distance that that part of the glacier had moved during the elapsed time. Table 1 lists the corresponding rates of flow for the two sets of estimates. The units are "meters per day."

Table 1
Flow Rates Estimated for the Hoseason Glacier
(meters per day)

Three Year Span (x_i)	Five-Year Span (y_i)
.73	.72
.76	.74
.75	.74
.77	.72
.73	.72
.75	
.74	

Analysis

Let σ_X and σ_Y denote the true standard deviations associated with three-year estimates and five-year estimates, respectively. The hypotheses that we want to test are

$$H: \quad \sigma_X = \sigma_Y$$

vs.

$$A: \quad \sigma_X \neq \sigma_Y$$

For mathematical reasons, the statistic for deciding between H and A has a different form (and distribution) than the one we have used for problems involving location. Instead of focusing on the difference, $s_X^2 - s_Y^2$, we will look at the quotient, s_X^2/s_Y^2. If H is true, s_X^2/s_Y^2 should be close to 1. Therefore, we should reject the null hypothesis, in favor of a two-sided alternative, if s_X^2/s_Y^2 is either too much less than 1 or too much greater than 1.

The behavior of s_X^2/s_Y^2 is described by a family of curves known as the F distribution. These curves are similar to the χ^2 distribution except that they are indexed by *two* parameters, N and D. Both N and D are referred to as degrees of freedom, with N being equal to $n - 1$ and D being equal to $m - 1$.

In this particular example, we would say that the behavior of s_X^2/s_Y^2 is described by an F distribution with 6 and 4 degrees of freedom.

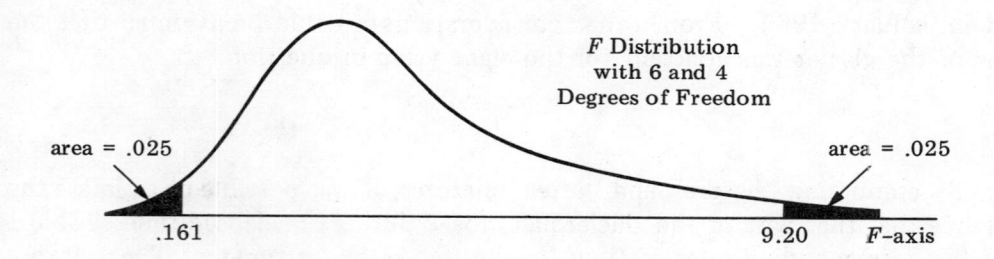

Since $P(F \leqslant .161) = .025$ and $P(F \geqslant 0.20) = .025$,

$$P(.161 < s_X^2/s_Y^2 < 9.20) = .95$$

Therefore, in testing

$$H: \quad \sigma_X = \sigma_Y$$

vs.

$$A: \quad \sigma_X \neq \sigma_Y$$

H should be rejected at the $P = .05$ level of significance if s_X^2/s_Y^2 is either

1. less than or equal to .161

or

2. greater than or equal to 9.20.

From Table 1,
$$\sum_{i=1}^{7} x_i = 5.23$$

and

$$\sum_{i=1}^{7} x_i^2 = 3.9089$$

so that

$$s_X^2 = \frac{7(3.9089) - (5.23)^2}{7(6)} = \frac{.0094}{42}$$

$$= .000224$$

Similarly,

$$\sum_{i=1}^{5} y_i = 3.64$$

and

$$\sum_{i=1}^{5} y_i^2 = 2.6504$$

making

$$s_Y^2 = \frac{5(2.6505) - (3.64)^2}{5(4)} = \frac{.0024}{20}$$

$$= .000120$$

Therefore, the ratio of sample variances is

$$\frac{s_X^2}{s_Y^2} = \frac{.000224}{.000120}$$

$$= 1.87$$

Question 3.6.1

What conclusion do we reach regarding H and A?

Question 3.6.2

Why was the alternative hypothesis for this problem two sided rather than one sided?

Question 3.6.3

Test

$$H: \sigma_X = \sigma_Y$$

vs.

$$A: \sigma_X \neq \sigma_Y$$

for the Twain–Snodgrass data of Example 3.2. Use the $P = .05$ level of significance. Would variability be a good criterion to use in disputed authorship problems?

Question 3.6.4

Note that the left-hand and right-hand critical values for testing

$$H: \sigma_X = \sigma_Y$$

vs.

$$A: \sigma_X \neq \sigma_Y$$

both converge to 1 as n and m increase. Why should this be so?

REFERENCE

1. Morgan, Peter J. "A Photogrammetric Survey of Hoseason Glacier, Kemp Coast, Antarctica" (*Journal of Glaciology*, 12, 1973, pp. 113-120).

3.3 A two-sample test for proportions (Examples 3.7-3.8)

In Section 3.2, the two random samples $(x_1, x_2, \ldots, x_n)$ and $(y_1, y_2, \ldots, y_m)$ were assumed to be continuous measurements. Not every two-sample problem, though, falls into that same format. Suppose, for example, a physician is comparing two different treatments for the same condition. And suppose that n patients are given the first treatment and m, the second, and that the measured response is either "yes, the treatment worked" or "no, the treatment did not work." Let x_i denote the outcome for the ith patient given the first treatment and y_i, the outcome for the ith patient given the second treatment. Then $(x_1, x_2, \ldots, x_n)$ and $(y_1, y_2, \ldots, y_m)$ represent the outcomes of two independent series of binomial trials.

The parameters of interest when the two samples are binomial are p_X, the probability of a success with the first treatment, and p_Y, the probability of success with the second treatment. The hypotheses that we want to test are

$$H: p_X = p_Y$$

vs.

$$A: p_X \neq p_Y \text{ (or } A: p_X < p_Y \text{ or } A: p_X > p_Y\text{)}$$

The appropriate test statistic is just an extension of the Z-transformation that was used for one-sample problems. If H is true, the behavior of

$$\frac{(x/n) - (y/m)}{\sqrt{\dfrac{[(x+y)/(n+m)][1-(x+y)/(n+m)]}{n+m}}}$$

is approximated by the standard normal curve, where x and y are the numbers of successes in the two sets of data. If the alternative was two sided, and if the level of significance was $P = .05$, values of

$$\frac{(x/n) - (y/m)}{\sqrt{\dfrac{[(x+y)/(n+m)][1-(x+y)/(n+m)]}{n+m}}}$$

less than or equal to -1.96 or greater than or equal to $+1.96$ would require that the null hypothesis be rejected. (Why?)

Example 3.7 Bumper stickers

Law and order was a key issue for all the presidential contenders in 1968 but particularly so for George Wallace. He and his supporters were the sharpest critics of the courts and the strongest advocates of a renewed commitment to vigorous law enforcement. But whenever one segment of society moralizes to another, there is a natural tendency for the latter to question whether the former are, themselves, above reproach. "Wallaceites" talked a good game of law and order, but did they practice what they preached? In Nashville, Tennessee, a team of political scientists carried out a rather unorthodox survey that seemed to indicate that they didn't.

This is a good example for illustrating the point that the application of statistical techniques is not limited to laboratory situations or very technical experiments. By using a little imagination, researchers in almost any field can put the basic principles of statistical inference to very good use.

Objective

To see whether supporters of a strong law and order candidate and supporters of a moderate law and order candidate differed in their own observance of the law.

Procedure

Prior to the general election of 1968, the government of Nashville–Davidson County had passed a law requiring all locally operated vehicles to display a "Metro sticker" on their windshields. The cost per car was $15.00. It was not entirely clear, though, what the penalty would be for not having a sticker or how strictly the law would be enforced.

For several days following the sticker deadline (November 1), spot surveys were made of various parking lots in Nashville to see whether supporters of Humphrey and Wallace differed significantly in their compliance with the new law. Table 1 shows some of the results.

Data

Table 1
Parking Lot Surveys of Bumper Stickers

In support of*	Number of cars	Number with stickers
Humphrey	$n_H = 178$	$X_H = 154$
Wallace	$n_W = 361$	$X_W = 270$

*A car was assumed to be owned by a Humphrey supporter, for example, if it displayed a Humphrey bumper sticker.

Analysis

The sample proportions of Humphrey and Wallace supporters obeying the law were 154/178 = .865 and 270/361 = .748, respectively. Presumably, these two figures are unbiased estimates of p_H and p_W, the true proportions of Humphrey and Wallace cars bearing a Metro sticker. Can we conclude that p_H and p_W are not equal, as the two sample proportions would indicate, or is the observed difference of .117 (= .865 - .748) well within the normal bounds of sampling variability? Put more formally, the hypotheses to be tested are

$$H: p_H = p_W \quad (= \hat{p})$$

vs.

$$A: p_H \neq p_W$$

If H is true, a pooled estimate of p, denoted by the symbol $\hat{p}$, would be the overall sample proportion of cars with stickers. That is,

$$\hat{p} = \frac{154 + 270}{178 + 361} = \frac{424}{539} = .787$$

Also, recall that the sampling distribution of

$$\frac{\dfrac{X_H}{n_H} - \dfrac{X_W}{n_W}}{\sqrt{\dfrac{\hat{p}(1-\hat{p})}{n_H} + n_W}}$$

is described by the standard normal curve. Therefore,

$$P\left\{-2.58 < \frac{\dfrac{X_H}{n_H} - \dfrac{X_W}{n_W}}{\sqrt{\dfrac{\hat{p}(1-\hat{p})}{n_H} + n_W}} < 2.58\right\} = .99$$

It follows that the null hypothesis should be rejected at the $P = .01$ level of significance if

$$\frac{\dfrac{X_H}{n_H} - \dfrac{X_W}{n_W}}{\sqrt{\dfrac{\hat{p}(1-\hat{p})}{n_H} + n_W}}$$

is either

1. less than or equal to -2.58

or

2. greater than or equal to $+2.58$.

Since $X_H/n_H = .865$, $X_W/n_W = .748$, and $\hat{p} = .787$, the test statistic is equal to

$$\frac{.865 - .748}{\sqrt{\dfrac{(.787)(.213)}{178} + 361}} = 6.65$$

Our conclusion is obvious: we should reject the null hypothesis. The proportion of Wallace supporters breaking the law by not having a Metro sticker was significantly higher than the corresponding proportion for Humphrey supporters!

Question 3.7.1

At least two explanations might be offered for such a disparity in compliance rates. On the one hand we could contend that Wallace supporters (more so than Humphrey supporters) see themselves as being *above* the law—to the extent that their concern for justice stops somewhere short of local tax ordinances. On the other hand, the effect being observed could be an economic one. Gallup polls have shown that a relatively greater percentage of Wallace supporters come from the lower socioeconomic classes. Maybe they could not afford the sticker. Table 2 shows the compliance rates for Wallace supporters and for Wallace "controls." (A Wallace control was a vehicle without any political sticker parked next to a vehicle with a Wallace sticker. Presumably, the owners of Wallace cars and Wallace control cars would be in similar socioeconomic classes, by virtue of their cars being in the same parking lot.

Table 2
Compliance Rates for Wallace Cars and Wallace Controls

Classification	Number of cars	Number with stickers
Wallace cars	361	270
Wallace controls	361	303

What do you conclude from Table 2?

Question 3.7.2

What other information might be collected on a survey of this sort to help resolve the socioeconomic question?

Question 3.7.3

Construct a 99 per cent confidence interval for $p_H - p_W$ using the data of Table 1.

Question 3.7.4

Suppose the true porportion of Wallace cars having Metro stickers is .75. What is the probability that 80 or more of the next 100 Wallace cars spotted will have Metro stickers? Estimate the probability that *exactly* 80 will have Metro stickers?

REFERENCE

1. Wrightsman, Lawrence S., "Wallace Supporters and Adherence to 'Law and Order,'" in *Human Social Behavior*, Robert A. Baron and Robert M. Liebert, eds. (Dorsey Press, 1971, pp. 217-225).

Example 3.8 Nightmare sufferers

We have seen that most data fall into two broad categories, binomial and continuous. The distinction is important because the procedures that are appropriate for one may not be appropriate for the other. There is also a distinction between the two with respect to their origins. Continuous data are much more common in the natural sciences; binomial data are much more common in the social sciences. This example shows a typical two-sample, binomial data problem.

Many studies have sought to characterize the nightmare sufferer. Out of these has emerged the stereotype of someone with high anxiety, low ego strength, feelings of inadequacy, and poorer than average physical health. What is not so well known is whether this pattern is more typical of men or of women.

Statistics in the Real World

> **Objective**
>
> To test whether men have nightmares as often as women.

Procedure

A total of 352 persons were interviewed, 160 men and 196 women. Each was asked whether he or she had nightmares "often" (at least once a month) or "seldom" (less than once a month).

Data

Table 1

Nightmare Frequency	Men	Women
Often	55	60
Seldom	105	132

Question 3.8.1

Define the parameters in this problem and state the appropriate null and alternative hypotheses.

Question 3.8.2

If the null hypothesis of Question 3.8.1 is true, what would be the sample estimate of the proportion of people who have nightmares often?

Question 3.8.3

State the statistic appropriate for testing H versus A, as defined in Question 3.8.1. Give the distribution of the test statistic.

Question 3.8.4

Test the hypotheses of Question 3.8.1 at the $P = .05$ level of significance.

Question 3.8.5

Continuous data can always be transformed into binomial data by designating every measurement in a certain interval (or set of intervals) as a "success" and any measurement outside that interval (or set of intervals) as a "failure." Suppose the data from an experiment consisted of two relatively large sets of continuous measurements. How might those measurements be transformed into binomial data so that $H: p_X = p_Y$ could be tested in lieu of $H: \mu_X - \mu_Y$?

REFERENCE

1. Hersen, Michel, "Personality Characteristics of Nightmare Sufferers" (*Journal of Nervous and Mental Diseases*, 153, 1971, pp. 29-31).

Chance is always powerful. Let your hook be always cast; in the pool where you least expect it, there will be a fish.

Ovid

Chapter 4

The Paired-data Problem

4.1 Introduction

Too often experimenters think of statisticians as people to consult only *after* a study is finished and all the results are in. That's like calling a doctor when the patient is dead! There is much more that can be gotten from statistical methodology than a postmortem analysis of what has already happened. Equally important, for example, is the problem of *designing* experiments that will yield the sorts of information the researcher is looking for. In this chapter, we consider, for the first time, situations where the design is as important as the analysis.

When the effects of two treatments are to be compared, we often have our choice between setting the experiment up according to the two-sample framework of the previous chapter or in the paired-data format presented here. Which is better? That depends on the circumstances. One of the purposes of the examples in this chapter is to point out the kinds of situations for which the paired-data model is especially helpful.

But before taking up the paired-data model, we should take a closer look at the two-sample t test of Chapter 3—and, in particular, at its weaknesses. One of these weaknesses involves what is known as the "subject effect." When Treatment X is applied to, say, Subject 10, the measured response, x_{10}, reflects not only the effect of the treatment but also the effects of all those many peculiarities that make Subject 10 different from Subject 9, or Subject 13. If these subject effects are very pronounced, the treatment effect, which is what we are primarily interested in, may be obscured.

For example, consider the following two experiments, both involving five observations taken on each of two treatments, X and Y.

Experiment 1

Treatment X	Treatment Y
30	38
28	39
31	38
30	42
31	43

Experiment 2

Treatment X	Treatment Y
15	50
45	20
30	10
5	80
55	40

In each case, the average for the X observations is 30; for the Y observations, 40. In Experiment 1, there is so little variability *within* the treatments that the difference *between* the treatments—namely, $\bar{x} - \bar{y} = 40 - 30 = 10$—is clearly significant. But in Experiment 2, the "within" variability is so large that we can no longer attach much significance to the "between" difference of 10.

As indicated before, one reason for a large within-treatment variability might be the subjects themselves. Suppose, for example, that Treatments X and Y were two drugs for lowering blood pressure. If we gave the drugs only to patients whose medical conditions were similar to begin with—say, "overweight males, ages 50+—we would expect the responses to be fairly consistent, as in Experiment 1. On the other hand, if we tested the drugs on ten randomly selected patients, whatever their conditions, we could expect to get highly variable responses, as in Experiment 2.

There are several ways to cope with measurements that exhibit a strong subject effect. Theoretically, sample sizes can be increased to a point where $\sigma_X/\sqrt{n}$ and $\sigma_Y/\sqrt{m}$ are sufficiently small that any reasonable difference in μ_X and μ_Y will be detected with a high probability. But that point is often well beyond what the researcher's budget will allow. A more practical approach is to match up similar subjects by twos and assign different treatments to members of the same pair. That way many of the sources of variability unrelated to the treatment will have been kept constant (or nearly so), and the treatment effect, if there is one, will stand out more clearly. Figure 1 illustrates this very important conceptual difference between the two-sample and paired-data models.

Figure 1

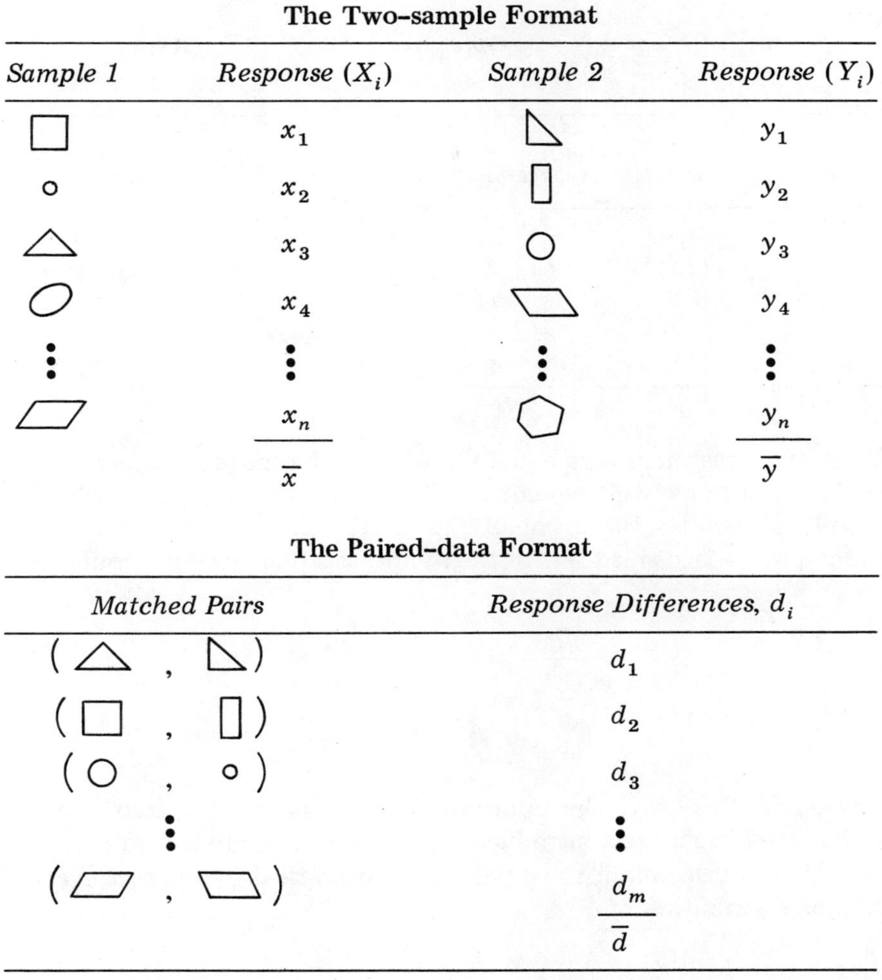

The Two-sample Format

Sample 1	Response (X_i)	Sample 2	Response (Y_i)
□	x_1	◿	y_1
○	x_2	▯	y_2
△	x_3	○	y_3
⬭	x_4	▱	y_4
⋮	⋮	⋮	⋮
▱	x_n	⬠	y_n
	$\overline{x}$		$\overline{y}$

The Paired-data Format

Matched Pairs	Response Differences, d_i
(△ , ◿)	d_1
(□ , ▯)	d_2
(○ , ○)	d_3
⋮	⋮
(▱ , ▱)	d_m
	$\overline{d}$

As you read the examples in this chapter, put yourself in the experimenters' shoes. Try to see what there was about the experimental conditions that led them to adopt the paired-data format.

4.2 The paired t test (Examples 4.1-4.7)

Even though the paired-data model involves two treatment groups, its analysis proceeds like the one-sample problems we did earlier. This is because we replace the original two sets of observations with a single set of differences—and then work with only those differences. Suppose, for example, a set of 10 subjects has been paired according to some criterion to form the five groups shown in Table 1. To one member of the ith pair (randomly chosen) we apply Treatment

X—and record x_i, to the other member we apply Treatment Y, and observe y_i. Then the response difference, $d_i = y_i - x_i$, is computed for each pair.

Table 1
The Paired-data Format

Pair Number	Subjects	Treatment X	Treatment Y	$d_i = y_i - x_i$
1	(4,8)	x_1	y_1	$d_1 = y_1 - x_1$
2	(2,1)	x_2	y_2	$d_2 = y_2 - x_2$
3	(5,2)	x_3	y_3	$d_3 = y_3 - x_3$
4	(7,10)	x_4	y_4	$d_4 = y_4 - x_4$
5	(3,9)	x_5	y_5	$d_5 = y_5 - x_5$

If the two treatments are equally effective, the means (μ_X and μ_Y) of their response distributions will be equal. Equivalently, $\mu_Y - \mu_X$ will be 0. If $\mu_d = \mu_Y - \mu_X$ denotes the mean of the distribution of response differences, it follows that, say, a two-sided test of the hypothesis that $\mu_X = \mu_Y$ could be written

$$H: \mu_d = 0$$

vs.

$$A: \mu_d \neq 0$$

Phrased in this way, the appropriate analysis for a paired-data model is simply the one-sample t test introduced in Chapter 2. Only the notation has been changed. If H is true, and if the n pairs have response differences $d_1, d_2 \ldots, d_n$, the sampling distribution of

$$\frac{\bar{d}}{s_d / \sqrt{n}}$$

is approximated by a Student t curve with $n - 1$ degrees of freedom, where

$$s_d = \sqrt{\frac{n \sum_{i=1}^{n} d_i^2 - (\sum_{i=1}^{n} d_i)^2}{n(n-1)}}$$

Values of $\bar{d}/(s_d/\sqrt{n})$ that are too much smaller than 0 or too much larger than 0 will result in $H: \mu_d = 0$ being rejected in favor of $A: \mu_d \neq 0$.

Example 4.1 "In lane number one..."

Paired experiments arise in many different contexts. In medical research, pairing is often done on the basis of age, sex, and overall physical condition. A common practice in animal studies is to form pairs out of littermates. And, of course, any data collected on a "before" and "after" basis is automatically paired.

For the behavioral study described here still another criterion was used. The subjects of this experiment were rats, and they were eventually paired according to how fast they could swim. This could hardly be considered a typical criterion, but under the circumstances it seemed like a reasonable thing to do.

Objective

To see what effect a rat's environment early in life has on its behavior late in life—in particular, do rats reared in isolation react differently in a competitive-stress situation than rats raised among siblings?

Procedure

Forty-six rats, all coming from one of several litters, were separated from their mothers when they were three weeks old. One half (Group A) were put into individual cages; the other half (Group B) were left together. After living under these conditions for about eight months, each rat was put through a series of "time trials" to determine its underwater swimming speed. Each trial involved only a single rat. (Speed was expressed in terms of the time required to swim a specified distance.) The average swimming time, $\bar{x}$, and the standard deviation of the swimming times, s, were calculated for each animal.

Pairings were made on the basis of these results. A rat a, belonging to Group A, was matched with a rat b, belonging to Group B, if either $\bar{x}_b$ was in the interval $\bar{x}_a \pm s_a$ or if $\bar{x}_a$ was in the interval $\bar{x}_b \pm s_b$. That is, the average swimming speeds of members of a pair differed by no more than one standard deviation.

Then, pair by pair, the rats were thrust into a competitive situation. Both members of a matched pair were submerged at the right-hand end of the tank illustrated in Figure 1. The only way they could get air was to swim the length of the tank and come out an escape hatch located near the other end.

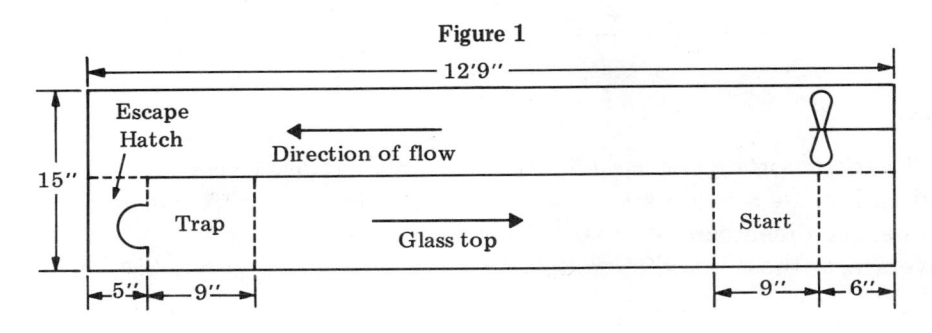

Figure 1

But only the *first* rat reaching the hatch was let out immediately—the other was kept underwater for a while longer, leaving no doubt as to who won the race! Altogether, each of the 23 pairs competed in 60 races.

Data

Table 1 shows the average times required for each animal to swim the length of the tank under the competitive situation just described. Each average is based on $n = 60$ races. The fifth column gives the difference, in average time, between members of each pair.

Table 1
Swimming Times in Competitive Trials

Matched Pairs Isolated	Normal	Average Times (sec) Isolated, x_i	Normal, y_i	Average Difference $d_i = y_i - x_i$
Rat Number 4	Rat Number 14	15.2	10.6	-4.6
6	30	12.4	13.7	1.3
16	7	9.1	11.7	2.6
24	31	9.2	8.2	-1.0
26	53	6.7	13.5	6.8
27	43	8.3	9.9	1.6
30	45	10.7	9.2	-1.5
31	6	10.0	12.4	2.4
35	47	13.1	11.7	-1.4
37	23	13.9	9.5	-4.4
45	2	10.1	7.9	-2.2
52	3	9.5	11.9	2.4
54	55	10.1	9.3	-0.8
55	32	6.6	8.7	2.1
17	25	8.1	7.7	-0.4
34	44	10.1	9.4	-0.7
41	40	9.7	8.9	-0.8
56	21	10.4	7.6	-2.8
21	20	8.1	7.6	-0.5
33	16	9.2	10.8	1.6
36	33	8.7	8.8	0.1
44	54	11.2	11.3	0.1
47	42	8.5	8.9	0.4

Analysis

If early environment has no effect on a rat's ability to cope with a competitive situation, we would expect there to be no consistent difference in swimming times between members of a pair. Rats raised alone should fare just as well, on the average, as those raised in groups. In other words, if $d_i (= y_i - x_i)$ denotes the difference in average times recorded for the members of the ith pair, we would expect the population mean of the d_i distribution, μ_d, to be 0. This suggests that we test

$$H: \mu_d = 0$$

vs.

$$A: \mu_d \neq 0.$$

If the null hypothesis is true, we would expect $\bar{d}$, the average of the d_i's, to be close to 0, since $\bar{d}$ is an unbiased estimate of μ_d. It follows that H should be rejected if $\bar{d}$ is either too much less than 0 or too much greater than 0.

Recall that the sampling distribution of

$$\frac{\bar{d}}{s_d/\sqrt{23}}$$

will be approximated by a Student t curve with 22 ($= n - 1$) degrees of freedom.

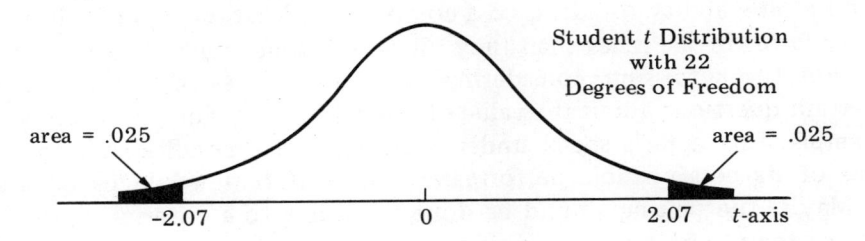

Since the numbers -2.07 and +2.07 cut off tail areas of .025 under this particular curve, we should reject H at the $P = .05$ level of significance if $\bar{d}/(s_d/\sqrt{23})$ is either

1. less than or equal to -2.07

or

2. greater than or equal to +2.07.

From Table 1,

$$\sum_{i=1}^{23} d_i = 0.3$$

and

$$\sum_{i=1}^{23} d_i^2 = 136.51$$

so that

$$\bar{d} = \frac{0.3}{23} = 0.01$$

and

$$s_d = \sqrt{\frac{23(136.51) - (0.3)^2}{23(22)}} = 2.49$$

Therefore,

$$\frac{\bar{d}}{s_d/\sqrt{23}} = \frac{0.01}{2.49/\sqrt{23}} = 0.02$$

Because 0.02 lies between −2.07 and +2.07, we accept the null hypothesis.

Discussion

Two things were clear from this experiment: (1) type of environment had no effect on a rat's ability to adjust to a competitive situation, and (2) all rats swam markedly faster in the races than they did in the time trials. This latter observation is not too surprising—considering the consequences of losing—but it does raise certain questions about the value of the pairing criterion. How do we know, for example, that a rat's speed under noncompetitive conditions is an accurate measure of its best possible performance—the sort that would be elicited in a race? Maybe the pairing should be done according to a criterion more directly relevant to the ultimate test conditions.

On a more fundamental note, can we be sure that the very nature of an underwater race—one that takes on all the appearances of a life-and-death struggle—does not override the behavioral differences we are looking for? Perhaps the variable being measured is not capable of answering the question being asked.

Question 4.1.1

Phrase the $P = .05$ decision rule in terms of $\bar{d}$.

Question 4.1.2

What other sort of information, besides the difference in average speed, might be used to compare the swimming performances of members of a pair?

Question 4.1.3

Suppose the following eight people had volunteered to participate in (a) a weight-reducing experiment and (b) a vitamin deficiency experiment. If both studies were to be done in the paired-data format, how would you define the pairs in each case?

Subject	Sex	Age	Weight	Height	Occupation	Overall Health
A. T.	M	26	185	6'0"	policeman	very good
H. W.	M	42	180	5'6"	junior executive	fair
B. P.	F	38	95	5'5"	doctor	fair
C. S.	M	48	202	5'10"	construction worker	poor
M. R.	F	60	165	5'7"	housewife	good
J. J.	F	58	125	5'5"	teacher	very good
K. T.	F	44	100	5'7"	law clerk	poor
C. K.	M	29	180	5'7"	farmer	excellent

REFERENCE

1. Bayroff, A. G., "The Experimental Social Behavior of Animals II. The Effect of Early Isolation of White Rats on Their Competition in Swimming," in *Animal Social Psychology*, Robert B. Zajonc, ed. (Wiley, 1969, pp. 269-277).

Example 4.2   ESP and hypnosis

Probably the most common paired data problems are the ones in which the subjects are self paired. That is, each subject is measured twice, once "before" the treatment is applied and, again, "after" the treatment is applied. Statistically, this means that each subject is being paired with itself and acts as its own control. The study described here is an excellent example of the effective use of this experimental design.

Research in extrasensory perception (ESP) has taken many different directions over the years. Recently, considerable attention has been given to the possibility that hypnosis may be helpful in bringing out ESP in persons who did not think they had any. The obvious way to test such a hypothesis is to compare the ESP abilities of subjects when they are awake ("before") to when they are hypnotized ("after").

Objective

To determine whether a person's ability to guess cards improves under hypnosis.

Procedure

ESP experiments are often done with Zener cards. These are special cards having only five denominations:

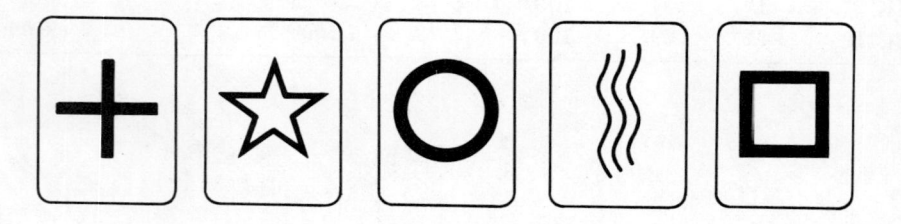

In this particular study, fifteen college students were each asked to guess the identity of 200 Zener cards. The same "sender"—that is, the person concentrating on the card—was used for each trial. For 100 of the trials, both the student and the sender were awake; for the other 100, both were hypnotized.

Data

Table 1 lists the number of correct identifications made by each of the subjects under both conditions. If chance were the only factor involved, the expected number of correct identifications in each 100 trials would be 20 (see Question 4.2.3).

Table 1
Number of Correct Responses (out of 100)

Student	Sender and Student in Waking State	Sender and Student in Hypnotic State
1	18	25
2	19	20
3	16	26
4	21	26
5	16	20
6	20	23
7	20	14
8	14	18
9	11	18
10	22	20
11	19	22
12	29	27
13	16	19
14	27	27
15	15	21

Analysis

Let x_i and y_i denote the scores of the ith student when awake and when hypnotized, respectively. The treatment differences, $d_i = y_i - x_i$, and their squares are shown in the last two columns of Table 2.

Table 2

Student	Score When Awake (x_i)	Score When Hypnotized (y_i)	$d_i = y_i - x_i$	d_i^2
1	18	25	7	49
2	19	20	1	1
3	16	26	10	100
4	21	26	5	25
5	16	20	4	16
6	20	23	3	9
7	20	14	-6	36
8	14	18	4	16
9	11	18	7	49
10	22	20	-2	4
11	19	22	3	9
12	29	27	-2	4
13	16	19	3	9
14	27	27	0	0
15	15	21	6	36
			43	363

Using the totals of the last two columns,

$$\bar{d} = \frac{43}{15} = 2.9$$

and

$$s_d = \sqrt{\frac{15(363) - (43)^2}{15(14)}} = \sqrt{17.12} = 4.1$$

Assuming hypnosis would not tend to *inhibit* a person's ESP ability, it would be reasonable in this situation to test the null hypothesis against a one-sided alternative. That is,

$$H: \mu_d = 0$$

vs.

$$A: \mu_d > 0$$

Since the sample size is 15, the distribution of

$$\frac{\bar{d}}{s_d/\sqrt{15}}$$

is described by a Student t curve with 14 degrees of freedom.

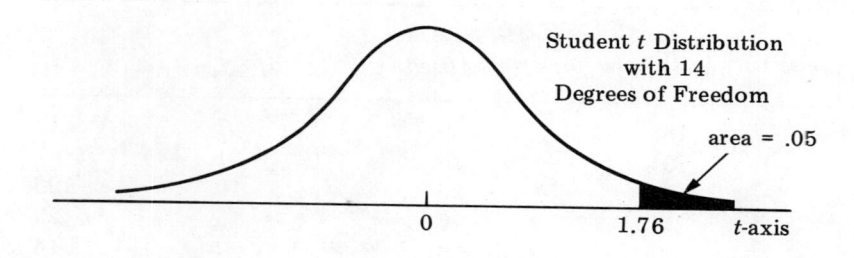

Student t Distribution with 14 Degrees of Freedom

area = .05

0   1.76 t-axis

If we choose to test H versus A at the $P = .05$ level of significance, we should reject the null hypothesis if $\bar{d}/(s_d/\sqrt{15})$ is greater than or equal to 1.76. But

$$\frac{\bar{d}}{s_d/\sqrt{15}} = \frac{2.9}{4.1/\sqrt{15}} = 2.74$$

Thus, our decision is to reject the null hypothesis and conclude that hypnosis *does* have an effect.

Question 4.2.1

Test H versus A at the $P = .01$ level of significance. At the $P = .005$ level of significance.

Question 4.2.2

How could this study be done as a two-sample problem? Which format, paired-data or two-sample, seems preferable? Why?

Question 4.2.3

If a subject is only guessing and has no ESP ability, his probability (p) of making the correct identification for any given card is $\frac{1}{5}$, or .20. From Table 1, the average number of correct guesses for the 15 hypnotized subjects was 21.7. Does this constitute substantial proof that ordinary people, when hypnotized, possess ESP ability?

160 Statistics in the Real World

REFERENCE

1. Casler, Lawrence, "The Effects of Hypnosis on GESP" (*Journal of Parapsychology*, 28, 1964, pp. 126–134).

Example 4.3 Glaucoma and cornea thickness

It was mentioned in Example 4.2 that when the subjects in an experiment are self paired, the measurements are recorded on a before and after basis. This is usually true, but not always true. Suppose, for example, a pharmaceutical house is developing two different flu vaccines (A and B) and wants to compare the localized side effects (that is, the amount of erythema) each one produces. One very good way would be to inject each subject twice—Vaccine A in one arm and Vaccine B in the other. If this were done, the subjects would certainly be self paired, but the measurements could not be considered "before" and "after."

The experiment described in this example is another instance of self-pairing that does not involve a before and after response. Here the purpose was to determine whether persons afflicted with glaucoma tended to have other eye defects. Specifically, do they have abnormally thick corneas?

At first it might seem as if the two-sample format would be called for: Sample 1 consisting of persons *without* glaucoma and Sample 2, persons *with* glaucoma. But sometimes people have unilateral glaucoma—only one eye being affected. By considering the two eyes of persons with this condition as a paired sample, we can examine the glaucoma-cornea relationship unobscured by subject-to-subject variability.

Objective

To determine whether glaucoma can be linked with another abnormality of the eye, increased cornea thickness.

Procedure

The sample subjects were eight persons with nonoperated unilateral glaucoma. Each was instructed to not take any treatment on the day of the examination. (The usual treatment for this condition would be medicine for reducing the increase in intraocular pressure (IOP) that characterizes glaucoma. But since that pressure is precisely what may cause a cornea to become abnormal, it was necessary to let the IOP rise unchecked.)

Data

Table 1 shows the thicknesses, measured in microns (μ), of the corneas of the eight subjects.

Table 1
Cornea Thickness (μ)

Patient	Sex	Glaucomatous Eye (High IOP)	Contralateral Eye (Low IOP)
K. H.	F	488	484
E. L.	M	478	478
M. J.	F	480	492
E. M.	F	426	444
K. F.	M	440	436
C. M.	F	410	398
A. T.	F	458	464
T. J.	F	460	476

Question 4.3.1

Define the parameters involved in this experiment and state the appropriate null and alternative hypotheses.

Question 4.3.2

Compute $\bar{d}$ and s_d. What distribution describes the behavior of $\bar{d}/(s_d/\sqrt{8})$?

Question 4.3.3

Test the hypotheses stated in Question 4.3.1 using the information computed in Question 4.3.2. Use the $P = .05$ level of significance.

Question 4.3.4

Construct a 95 per cent confidence interval for μ_d.

Question 4.3.5

Under what conditions would it be better to set an experiment up as a two-sample problem rather than a paired-data problem?

REFERENCE

1. Ehlers, Niels, "On Corneal Thickness and Intraocular Pressure, II" (*Acta Ophthalmologica*, 48, pp. 1107–1112).

Example 4.4 Bee stings

There are many factors that predispose a bee to sting. A person wearing dark clothing, for example, is more likely to get stung than someone wearing light clothing. And someone whose movements are quick and jerky runs a higher risk than a person who moves more slowly. Still another factor—one particularly important to beekeepers—is whether the person has just been stung by another bee.

This latter factor was simulated in a recent experiment. For obvious reasons, real beekeepers were not used—they were replaced by cotton balls wrapped in muslin.

Objective

To test whether honeybees show a preference for stinging objects that have already been stung.

Procedure

Eight cotton balls wrapped in muslin were dangled up and down in front of the entrance to a beehive. Four of the balls had just been exposed to a swarm of angry bees and were filled with stingers; the other four were fresh. After a specified length of time, the number of new stingers in each of the balls was counted. Eventually, the entire procedure was replicated eight more times.

Table 1
Number of Times Stung

Trial	Cotton Balls with Stings Already Present	Fresh Cotton Balls
1	27	33
2	9	9
3	33	21
4	33	15
5	4	6
6	21	16
7	20	19
8	33	15
9	70	10

Analysis

Most likely, the disposition of the bees at any given time will be subject to a variety of more or less transient factors, in addition to the cotton balls. As a result, the overall experimental environment may be quite different from trial to trial (see Question 4.4.3). Nevertheless, in any one trial all eight balls are tested under the same conditions, so the two measurements in each replication can be thought of as a pair.

Question 4.4.1

Analyze these data. Define all parameters. State the hypotheses being tested and the conclusion that is reached.

Question 4.4.2

How would this experiment be done using the two-sample format?

Question 4.4.3

Do the data bear out the suspicion that the overall experimental environment may be quite different from time to time? Explain.

REFERENCE

1. Free, J. B., "The Stimuli Releasing the Stinging Response of Honeybees" (*Animal Behavior*, 9, 1961, pp. 193–196).

Example 4.5 Drug therapy for learning problems

One last comment needs to be made with regard to designing a self-paired experiment: If there is the slightest chance that either of the treatments will have any residual effects, or systematic biases of any kind, the *order* in which the treatments are administered should be randomized. This problem failed to come up in either Example 4.2 or Example 4.3. In Example 4.2, the responses were before and after, so the order was already fixed. In Example 4.3, the measurements were very objective and beyond the influence of either the researcher or the subject.

Sometimes, though, particularly when treatments are given at different times and when the correct responses can be learned, order becomes a critical factor and must not be overlooked. The study described in this example falls into that category.

Children with severe learning problems often have electroencephalograms and behavior patterns similar to those of children with petit-mal, a mild form of epilepsy. This led to the speculation that whichever drugs were effective in treating petit-mal might also be useful as "learning facilitators." Recently a study was done that addressed itself to precisely this idea. The drug used was ethosuximide, a widely prescribed anticonvulsant.

Objective

To test whether drug therapy can improve the IQ of children with learning and behavioral problems.

Procedure

The subjects were 10 children, ranging in age from 8 to 14. The therapy was scheduled to last six weeks. For three of those weeks a child was given a placebo; for the other three weeks, ethosuximide. After each three week period the children were given several parts of the standard Wechsler IQ test. Because a child might be expected to do better on the test the second time he took it, the order in which the placebo and the ethosuximide were administered was randomized: Some children were given the placebo for the first three weeks while others began with the ethosuximide.

Data

Table 1 shows the two verbal IQ scores recorded for each subject. The order in which the treatments were given is not indicated.

Table 1
Verbal IQ Scores

Child	After Placebo (x_i)	After Ethosuximide (y_i)
1	97	113
2	106	113
3	106	101
4	95	119
5	102	111
6	111	122
7	115	121
8	104	106
9	90	110
10	96	126

Question 4.5.1

Graph the information presented in Table 1.

Question 4.5.2

Set up and test an appropriate hypothesis for these data. Use the $P = .01$ level of significance. Note: If $d_i = y_i - x_i$, then

$$\sum_{i=1}^{10} d_i = 120 \qquad \sum_{i=1}^{10} d_i^2 = 2,448$$

Question 4.5.3

What could be concluded if the "after ethosuximide" IQ's were significantly higher than the "after placebo" IQ's but if all 10 children had been given the placebo first?

Question 4.5.4

This experiment was done "double blind," meaning that neither the subjects nor the experimenters who were directly involved knew which treatment a child was being given at any particular time. Was that a necessary precaution? Explain.

REFERENCE

1. Smith, W. Lynn, "Facilitating Verbal-Symbolic Functions in Children with Learning Problems and 14-6 Positive Spike EEG Patterns with Ethosuximide (Zarontin)," in *Drugs and Cerebral Function*, Wallace Smith, ed. (Thomas, 1970, p. 125).

Example 4.6 Building a better mousetrap

To an applied statistician, what we have called the paired-data problem is really a special case of a more general structure known as a *randomized block design*. The term "block" refers to a complete replication of an experiment under conditions as similar as possible. When the experiment consists of only two treatments, a block is simply a pair.

Often a block is a replication done in a certain area or during a certain time. In this example, the experiment is replicated in five blocks, each done at a different place *and* a different time.

Objective

To test the effectiveness of artificial food flavoring in rat poison.

Procedure

Normally rat poison is made by mixing the active chemical ingredients with ordinary cornmeal. Unfortunately, in many urban areas rats can find food that they prefer to cornmeal. The result is that the poison is left untouched. One solution is to make the cornmeal more attractive by adding food supplements such as peanut butter or meat. This works but the cost is very high and the food supplements spoil rather quickly. The purpose of the experiment described in this example was to see whether artificial food supplements might have the same positive effects without any of the drawbacks.

Statistics in the Real World

The study took place in Milwaukee, Wisconsin, and consisted of five two-week surveys. For each survey approximately 1,600 baits were placed around garbage-storage areas; half of the baits were plain cornmeal, the other half were cornmeal mixed with artificial butter-vanilla flavoring. The baits were always placed in pairs so that any given rat would have equal access to both kinds.

After two weeks the sites were inspected and the number of baits that were gone was recorded. Then a different set of locations in the same general area were selected and the experiment was repeated for another two weeks. Altogether the study was replicated five times and lasted ten weeks.

Data

Listed in Table 1 are the bait-acceptance percentages for the two kinds of poison. In each case, the total number of baits placed was approximately 800.

Table 1
Percentage of Baits Accepted

Survey Number	Plain Flavor	Butter-Vanilla Flavor
1	.138	.117
2	.129	.167
3	.259	.298
4	.180	.231
5	.152	.202

Analysis

Since rats might actually dislike the butter-vanilla flavor, the null hypothesis should be tested against a two-sided alternative. Let x_i and y_i denote the acceptance percentages during the ith survey for the plain baits and the butter-vanilla baits, respectively. Let $d_i = y_i - x_i$. Let μ_d be the mean of the population distribution of d_i's. We want to test

$$H: \mu_d = 0$$

vs.

$$A: \mu_d \neq 0$$

Let $P = .05$. Because there are five differences, the distribution that describes the behavior of $\bar{d}/(s_d/\sqrt{5})$ is the Student t curve with 4 degrees of freedom.

The Paired-data Problem

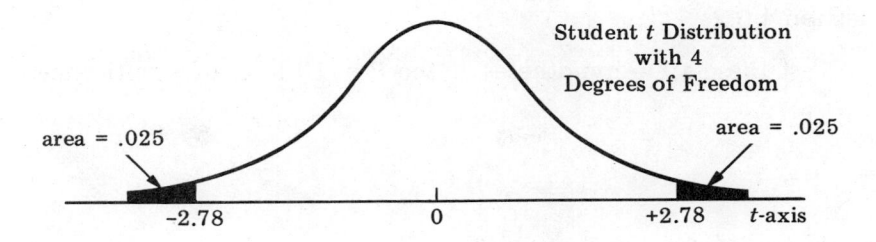

Student t Distribution with 4 Degrees of Freedom

area = .025   area = .025

−2.78   0   +2.78 t-axis

Since $P(-2.78 < t < 2.78) = .95$, H should be rejected if $\bar{d}/(s_d/\sqrt{5})$ is either (1) ≤ -2.78 or (2) $\geq +2.78$.

Referring to Table 1, we find that

$$\sum_{i=1}^{5} d_i = .157$$

and

$$\sum_{i=1}^{5} d_i^2 = .008507$$

Therefore,

$$\bar{d} = .0314$$

and

$$s_d = \sqrt{\frac{5(.008507) - (.157)^2}{5(4)}}$$

$$= \sqrt{.0008943} = .030$$

Question 4.6.1

What conclusion should be drawn?

Question 4.6.2

Test these same hypotheses at the $P = .10$ level of significance.

Question 4.6.3

Describe the blocks in Examples 4.1 through 4.5.

REFERENCE

1. Hulbert, Roger H., and Edward R. Krumbiegel, "Synthetic Flavors Improve Acceptance of Anticoagulant-Type Rodenticides" (*Journal of Environmental Health*, 34, 1972, pp. 407–411).

Example 4.7 How to stop smoking

The term "block" as defined in Example 4.6 referred to a particular set of conditions under which an entire experiment was replicated—that is, done once. Historically, many of the first applications of statistics were in connection with agricultural experiments. In this setting, a block was often a particular plot of ground, throughout which factors such as soil composition, amount of light, drainage, and so on were relatively homogeneous. The *concept* of a block, though, is really much more general. People, for instance, can be blocks. Of course, in Examples 4.2, 4.3, and 4.5, where the subjects were self-paired, the people were automatically blocks. But, in some cases, people can act as blocks even though they are not, themselves, the experimental subjects.

In the following study, an attempt was made to evaluate the effectiveness of two antismoking therapies. A total of seven psychologists participated. Each had charge of four subjects, two in each of the two therapies. Since each psychologist replicated the experiment—that is, tested both treatments—each one could be thought of as a block.

Objective

To compare the effectiveness of aversion therapy and supportive counselling in getting people to stop smoking.

Procedure

A total of 28 subjects were recruited from a college population. All were heavy smokers (more than 20 cigarettes a day) and all professed the desire to break the habit. Two different therapies were tested. Each involved nine treatment sessions spaced over a period of six weeks.

The first method was based on *aversion therapy*. The subject was instructed to take out a cigarette, light it, and smoke it in his usual way. While he was doing this, the therapist would be giving him a series of electric shocks.

The second method involved *supportive counselling*. This was a purely verbal approach, similar to nondirective psychotherapy.

Data

After the nine treatment sessions were completed, each subject recorded his or her current daily cigarette consumption. This figure was then converted to a percentage of the subject's initial cigarette consumption. Table 1 shows the average percentages for the two subjects in each therapy—therapist combination.

Table 1
Average Cigarette Consumption (per cent of baseline level)

Psychologist	Aversive conditioning	Supportive counselling
1	59.5	125.0
2	14.0	59.0
3	72.0	43.0
4	27.5	0.0
5	80.5	50.0
6	17.5	18.5
7	83.0	66.0

Question 4.7.1

Just by looking at the data, what impressions do you get about the effects of the different therapies and therapists?

Question 4.7.2

Analyze these data.

Question 4.7.3

Does it matter whether the treatment differences are expressed as percentages or as proportions when the t test is applied? Explain.

REFERENCE

1. Koenig, Karl P., and John Masters, "Experimental Treatment of Habitual Smoking" (*Behaviour Research and Therapy*, 3, 1965, pp. 235-243).

'Tis fate that flings the dice
and as she flings
of Kings makes peasants
and of peasants Kings.

Dryden

Chapter 5

The Correlation Problem

5.1 Introduction

Of all the different inference models, the correlation problem is probably the most familiar to persons without any formal training in statistics. The reason for this is that in the past few years several well-publicized, health-related controversies have been, from a statistical standpoint, correlation problems. Among these were the studies showing that fluoridated water helps prevent tooth decay, that smoking is related to heart disease, and that various air and water pollutants can cause cancer. Example 5.1 describes a similar sort of problem that is very much in the news today: the biological hazards posed by low-level radiation coming from the by-products of nuclear power plants.

In general, a correlation problem arises when each sample subject is measured for both an X-trait and a Y-trait, with the purpose being to examine the relationship between the two. For example, it is well known that the percentage of naturally occurring fluoride in municipal water supplies varies considerably from place to place. Furthermore, dental surveys done in a number of cities have shown that a definite relationship exists between tooth decay and fluoride concentration: In cities having higher than normal fluoride concentrations, the percentage of children with no caries tends to be higher than normal, and vice versa. In other words, the amount of fluoride in drinking water is *correlated* with the prevalence of tooth decay among children.

The correlation problems in this chapter fall into two generic types: those where both the X- and Y-traits are measured on continuous scales and those where both X and Y are qualitative. The first type is the subject of Examples 5.1 through 5.4; the second type is illustrated by Examples 5.5 through 5.9. The problems in this latter group are known as χ^2 (chi square) problems. In some disciplines, problems of this sort come up more often than all the other inference models put together.

As you read through the examples in this chapter, note how the structure of correlation problems differs from the structure of two-sample problems and paired-data problems. It has already been mentioned that correlation problems involve two observations, x_i and y_i, taken on each subject. Another characteristic (in most cases) is that the units of x_i and y_i are not comparable. That is, the difference $x_i - y_i$ has no physical meaning. It would not make sense, for example, to subtract the percentage of children in a certain city having no caries (y_i) from the fluoride concentration of the water supply in that city (x_i). This latter condition is what distinguishes a correlation problem from either a paired-data problem or a two-sample problem.

5.2 The least-squares line and the sample correlation coefficient (Examples 5.1–5.4)

The data in a correlation problem consists of n pairs of points $(x_1, y_1), (x_2, y_2)$, ..., (x_n, y_n), where x_i and y_i are the two responses recorded for the ith subject. When X and Y are both continuous, there are two questions that we need to answer:

1. What is the *nature* of the relationship between X and Y?
2. How *strong* is the relationship between X and Y?

In answering the first question we will distinguish two basic types of relationships, *linear* and *nonlinear*. A linear (or straight-line) relationship is one in which the overall pattern in the n points can be adequately approximated by an equation of the form $y = a + bx$.

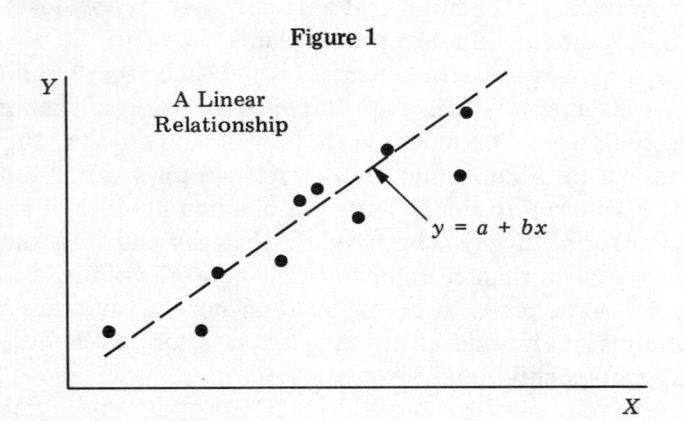

Figure 1

The constants a and b are called the *y-intercept* and *slope*. One of the procedures associated with data of this type is to find the "best" straight line that describes the given set of points. Of course, finding that line is equivalent to finding numerical values for a and b. According to the Least Squares Theorem,

$$b = \frac{n(\sum_{i=1}^{n} x_i y_i) - (\sum_{i=1}^{n} x_i)(\sum_{i=1}^{n} y_i)}{n(\sum_{i=1}^{n} x_i^2) - (\sum_{i=1}^{n} x_i)^2}$$

and

$$a = \frac{\sum_{i=1}^{n} y_i - b \sum_{i=1}^{n} x_i}{n}$$

Examples 5.1 and 5.2 explain in more detail how a and b are calculated.

Any relationship *not* of the form $y = a + bx$ is said to be *nonlinear*.

Figure 2

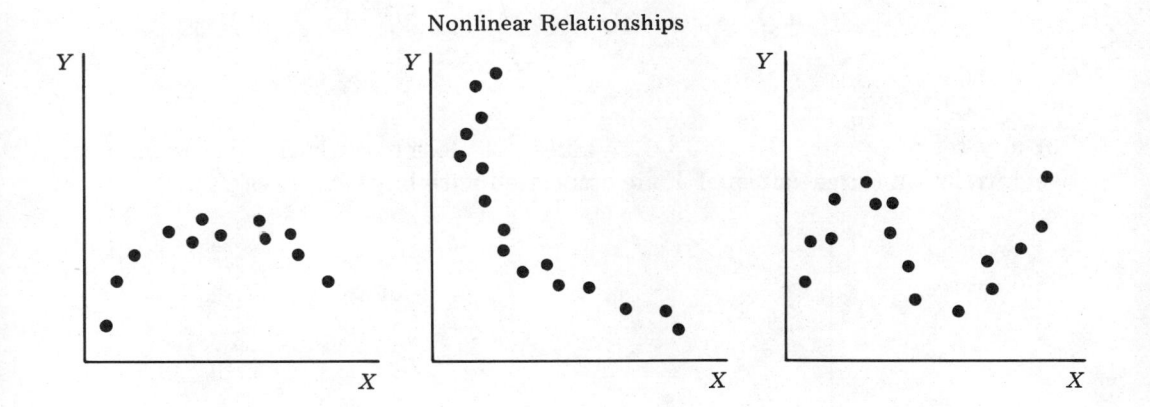

Nonlinear Relationships

The equations describing nonlinear relationships can take many different forms. They can be exponential ($y = ae^{bx}$), trigonometric ($y = a\sin x + b\cos x$), or polynomial ($y = a + bx + cx^2$), just to name a few. Although there are methods for analyzing data that are nonlinear, none will be considered in this chapter.

Knowing that a relationship is linear—even knowing the exact equation of the least squares line—does not tell us everything we need to know about X and Y. The two relationships graphed in Figure 3, for example, are both linear, but very different.

Figure 3

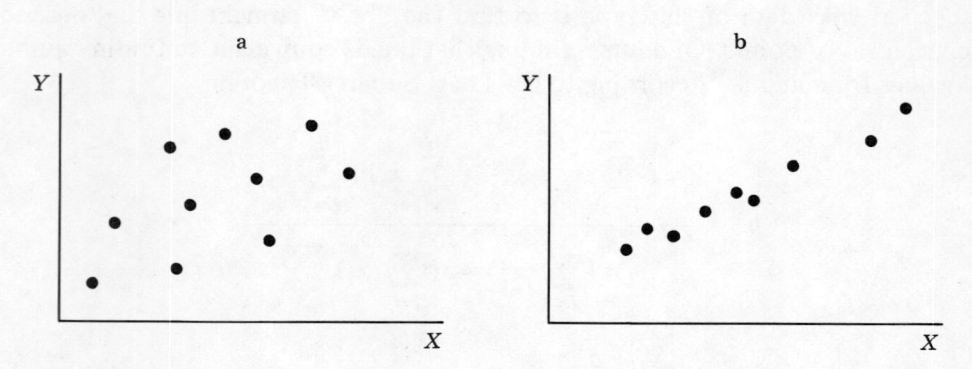

The linearity is much stronger in Figure 3b than it is in Figure 3a. This raises an obvious question. Is there some way that the strength of a linear relationship can be quantified? The answer is "yes" and it comes in the form of a statistic known as the *sample correlation coefficient*.

The sample correlation coefficient is denoted by the symbol r and is given by the expression

$$r = \frac{n \sum_{i=1}^{n} x_i y_i - (\sum_{i=1}^{n} x_i)(\sum_{i=1}^{n} y_i)}{\sqrt{n \sum_{i=1}^{n} x_i^2 - (\sum_{i=1}^{n} x_i)^2} \sqrt{n \sum_{i=1}^{n} y_i^2 - (\sum_{i=1}^{n} y_i)^2}}$$

For any set of points, $-1 \leq r \leq +1$. Values of r close to $+1$ indicate a strong positive correlation (large values of Y are associated with large values of X).

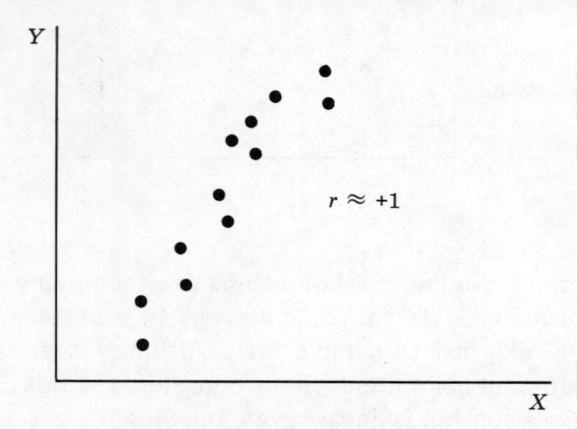

Values of r close to -1 indicate a strong negative correlation (large values of Y are associated with small values of X).

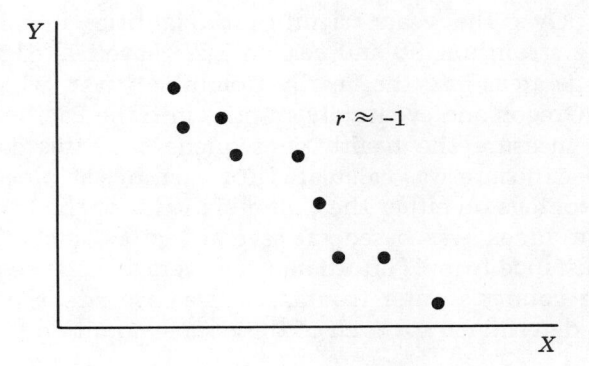

[Scatter plot showing Y vs X with points trending downward, labeled $r \approx -1$]

Values of r close to 0 indicate that there is essentially *no* linear relationship between X and Y.

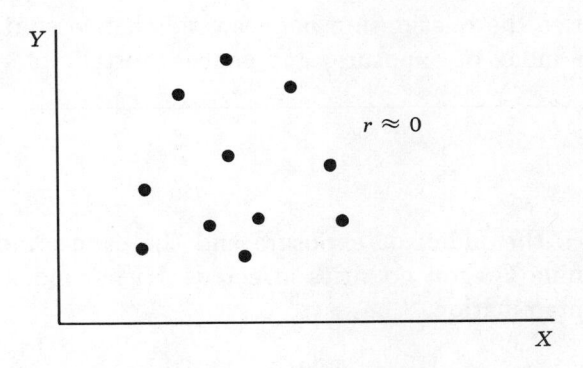

[Scatter plot showing Y vs X with scattered points, labeled $r \approx 0$]

Applications of the correlation coefficient are discussed in Examples 5.3 and 5.4.

Example 5.1 Radioactive contamination

The recent oil embargo has raised some very serious and disturbing questions about energy policies in the United States. One of the most controversial is whether nuclear reactors should assume a more central role in the production of electric power. Those in favor point to their efficiency and to the availability of nuclear material; those against warn of nuclear "incidents" and emphasize the health hazards posed by low-level radioactive contamination.

Since nuclear power is relatively new, there is not an abundance of past experience to draw on. One notable exception, though, is a government reactor that has been in continuous operation for 30 years. What happened there is what environmentalists fear will become commonplace if nuclear reactors are proliferated.

Since World War II, plutonium for use in atomic weapons has been produced at an AEC facility in Hanford, Washington. One of the major safety problems encountered there has been the storage of radioactive wastes. Over the years, significant quantities of these substances—including strontium 90 and cesium 137—have leaked from their open-pit storage areas into the nearby Columbia River, which flows through parts of Oregon and eventually empties into the Pacific Ocean.

To measure the health consequences of this contamination, an index of exposure was calculated for each of the nine Oregon counties having frontage on either the Columbia River or the Pacific Ocean. This particular index was based on several factors, including the county's stream distance from Hanford and the average distance of its population from the county's water frontage. As a covariate, the cancer mortality rate was determined for each of these same counties.

Objective

To summarize the relationship between radioactive contamination (as measured by the index of exposure) and cancer mortality.

Data

Table 1 shows the index of exposure and the cancer mortality rate (per 100,000) for the nine Oregon counties affected. Higher index values represent higher levels of contamination.

Table 1
Radioactive Contamination and Cancer Mortality in Oregon

County	Index of Exposure	Cancer Mortality per 100,000
Umatilla	2.49	147.1
Morrow	2.57	130.1
Gilliam	3.41	129.9
Sherman	1.25	113.5
Wasco	1.62	137.5
Hood River	3.83	162.3
Portland	11.64	207.5
Columbia	6.41	177.9
Clatsop	8.34	210.3

Analysis

The first step in summarizing the data of Table 1 is to plot the cancer mortality rate against the index of exposure for each of the nine counties.

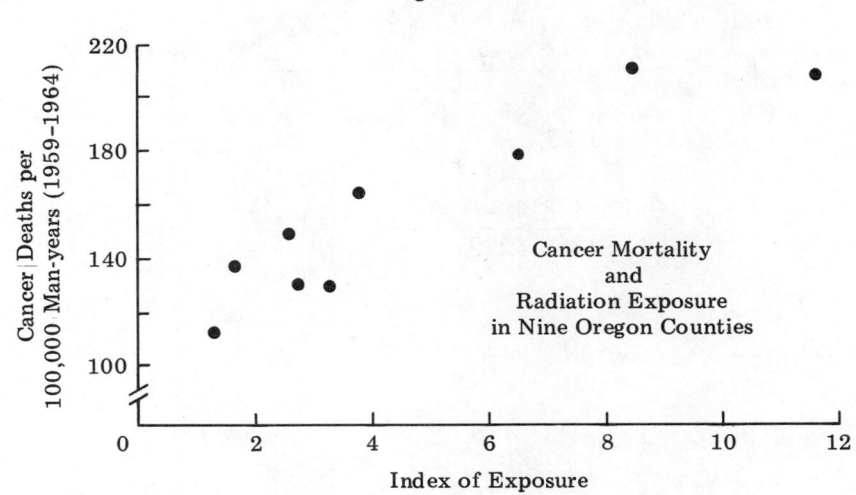

Figure 1

Cancer Mortality and Radiation Exposure in Nine Oregon Counties

(Scatterdiagram: Cancer Deaths per 100,000 Man-years (1959-1964) vs. Index of Exposure)

Graphs of this sort are known as *scatterdiagrams*. They are very useful for showing how two sets of measurements are related. In this case the relationship seems quite linear. Furthermore, the trend is "positive." Counties with high exposure indices tend to have high cancer mortality rates.

Having established that the data lie more or less on a straight line, the next step would be to find the equation of the *best* straight line that could be fit through the points. Suppose we let (x_i, y_i) denote the index of exposure and the cancer mortality rate for the ith county. Then

$$\sum_{i=1}^{9} x_i = 41.56 \qquad \sum_{i=1}^{9} x_i^2 = 289.4222$$

$$\sum_{i=1}^{9} y_i = 1{,}416.1 \qquad \sum_{i=1}^{9} y_i^2 = 232{,}498.97$$

$$\sum_{i=1}^{9} x_i y_i = 7{,}439.37$$

Using the Least Squares Theorem, the slope of the regression line is given by

$$b = \frac{n\sum_{i=1}^{n} x_i y_i - (\sum_{i=1}^{n} x_i)(\sum_{i=1}^{n} y_i)}{n\sum_{i=1}^{n} x_i^2 - (\sum_{i=1}^{n} x_i)^2}$$

$$= \frac{9(7{,}439.37) - (41.56)(1{,}416.1)}{9(289.4222) - (41.56)^2} = \frac{8{,}101.214}{877.5662}$$

$$= 9.23$$

Also, the y-intercept is given by

$$a = \frac{\sum_{i=1}^{n} y_i - b(\sum_{i=1}^{n} x_i)}{n}$$

$$= \frac{1{,}416.1 - (9.23)(41.56)}{9} = \frac{1{,}032.5012}{9}$$

$$= 114.72$$

Therefore, the line we are looking for has the equation

$$y = 114.72 + 9.23x$$

To plot the least squares line on the scatterdiagram, we simply need to choose two arbitrary values for x and solve the above equation for y. If $x = 2$, the predicted cancer mortality is

$$y = 114.72 + (9.23)(2) = 133.18$$

If $x = 10$,

$$y = 114.72 + (9.23)(10) = 207.02$$

When these points are connected, the regression line is determined (see Figure 2).

Figure 2

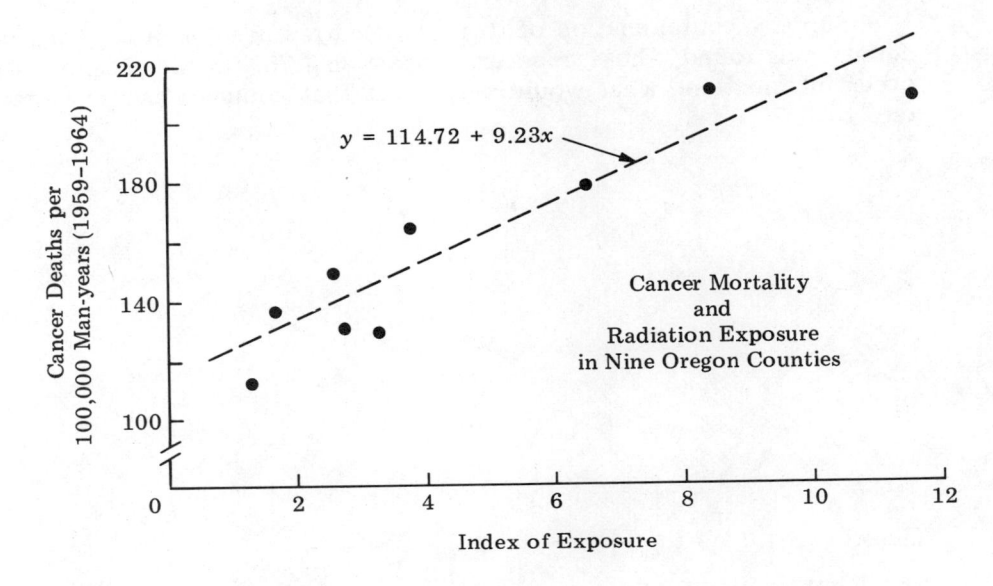

Cancer Deaths per 100,000 Man-years (1959-1964)

$y = 114.72 + 9.23x$

Cancer Mortality and Radiation Exposure in Nine Oregon Counties

Index of Exposure

Question 5.1.1

Wht is it important to plot data on a scatterdiagram before substituting the (x_i, y_i) values into the formulas for a and b?

Question 5.1.2

Every least squares straight line goes through the point $(\bar{x}, \bar{y})$. Verify that this is true for the line $y = 114.72 + 9.23x$.

Question 5.1.3

Suppose contamination of this sort occurred in some other place and a county was found whose exposure index was 4.75. In the absence of any other information, what would we expect that county's cancer mortality rate to be?

Question 5.1.4

In this analysis, exposure index (x) was plotted on the horizontal axis and cancer mortality rate (y), on the vertical axis. Suppose the two variables were interchanged, with mortality rate being plotted on the horizontal axis and exposure index on the vertical axis. If the Least Squares Theorem was applied to the data in *this* form, would the resulting regression line be the same as the one already found? Explain.

REFERENCE

1. Fadeley, Robert Cunningham, "Oregon Malignancy Pattern Physiographically Related to Hanford, Washington, Radioisotope Storage" (*Journal of Environmental Health*, 27, 1965, pp. 883-897).

Example 5.2 Calibrating a cricket

There are basically two reasons for calculating a least squares line. One is that the regression line provides a convenient way of summarizing the relationship between two sets of data, in much the same way that calculating $\bar{x}$ and s was a convenient way of summarizing the variability in a *single* set of data. The second is that the least squares line makes it easy to predict the value of y knowing only the value of x. This latter reason would be our motivation in the situation described below.

Crickets make their chirping sound by sliding one wing cover back and forth over the other, very rapidly. For a long time naturalists have been aware that there is a definite linear relationship between temperature and the frequency of the chirps. This implies that by knowing the mathematical equation that describes the chirping-temperature relationship, we can use a cricket as a thermometer. . . .

Objective

To calibrate the chirping of a cricket as a function of temperature.

Data

The precise nature of the temperature-chirping frequency relationship is not the same for all crickets. It always appears to be linear but the slope and y-intercept of the approximating least squares line varies from species to species. Listed below are 15 frequency-temperature observations recorded for the striped ground cricket, *Nemobius fasciatus fasciatus*.

Table 1
Chirping Frequency and Temperature

Observation Number	Chirps per sec (x_i)	Temperature, °F (y_i)
1	20.0	88.6
2	16.0	71.6
3	19.8	93.3
4	18.4	84.3
5	17.1	80.6
6	15.5	75.2
7	14.7	69.7
8	17.1	82.0
9	15.4	69.4
10	16.2	83.3
11	15.0	79.6
12	17.2	82.6
13	16.0	80.6
14	17.0	83.5
15	14.4	76.3

Question 5.2.1

Graph the data of Table 1. Plot the chirping frequency on the horizontal axis.

Question 5.2.2

Given that

$$\sum_{i=1}^{15} x_i = 249.8 \qquad \sum_{i=1}^{15} x_i^2 = 4{,}200.56$$

$$\sum_{i=1}^{15} y_i = 1{,}200.6 \qquad \sum_{i=1}^{15} x_i y_i = 20{,}127.47$$

find the equation of the least squares line that describes the chirping-temperature relationship. Plot the line on the graph drawn in Question 5.2.1.

Question 5.2.3

Suppose a cricket belonging to this species chirped 1,110 times in one minute. What would we estimate the temperature to be?

Question 5.2.4

What obvious drawback does *Nemobius fasiatus fasciatus* have as a thermometer?

REFERENCE

1. Pierce, George W., *The Songs of Insects* (Harvard University Press, 1949, pp. 12-21).

Example 5.3 Future shock

Finding a least squares line, $y = a + bx$, is an appropriate way to analyze a set of data if the corresponding scatterdiagram reveals an obvious linear relationship between the x- and y-variables. Sometimes, though, the data points are so scattered that we cannot be sure whether x and y follow *any* pattern, let alone a linear one. When this is the case, the statistical analysis should first establish whether x and y are *independent*. This is usually done by computing the sample correlation coefficient, r, and testing whether r is significantly different from 0. The following example is a case in point.

Not long ago, Alvin Toffler wrote a best-seller called *Future Shock* in which he speculated about the cultural, medical, and moral effects of change. One question that was raised involved the extent to which change should be considered a health hazard—if, indeed, it should be considered a health hazard at all. A partial answer was provided in a recent study done on a group of patients hospitalized for various chronic illnesses.

Objective

To test whether persons undergoing greater life-style changes develop more serious chronic illnesses.

Procedure

All of the patients included in this study had been hospitalized for chronic conditions, some serious and some not. Each patient was asked to fill out a Schedule of Recent Experience (SRE) questionnaire. The questions related to 42 different life-change situations (a new job, another child, and so forth). A composite score was computed that summarized the average degree of change each patient had experienced in the preceding two years. Higher values on the SRE questionnaire reflected greater life-style changes.

At the same time, each patient's health condition (the one he was being hospitalized for) was graded according to the Seriousness of Illness Rating Scale (SIRS). Values on this scale ranged from a low of 21 for dandruff to a high of 1020 for cancer.

Data

The patients in the sample had a total of 17 different conditions that were considered chronic. In some cases more than one patient had the same condition. Table 1 and Figure 1 show the SIRS values associated with the various illnesses as well as the average SRE scores recorded for the patients who had each one.

Table 1
Life-style Changes (SRE) and Health Evaluations (SIRS)

Admitting diagnosis	Average SRE	SIRS
dandruff	26	21
varicose veins	130	173
psoriasis	317	174
eczema	231	204
anemia	325	312
hyperthyroidism	816	393
gallstones	563	454
arthritis	312	468
peptic ulcer	603	500
high blood pressure	405	520
diabetes	599	621
emphysema	357	636
alcoholism	688	688
cirrhosis	443	733
schizophrenia	609	776
heart failure	772	824
cancer	777	1,020

Figure 1

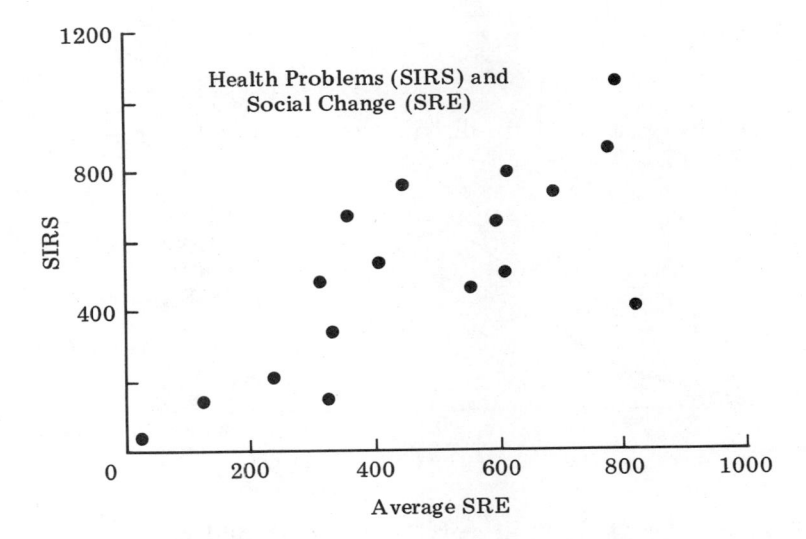

Health Problems (SIRS) and Social Change (SRE)

Analysis

The scatterdiagram for Table 1 suggests a possible relationship between SRE and SIRS but the linearity is certainly not as strong as it was in, say, Example 5.1.

Consequently, before fitting a least squares line through these 17 points it would be a good idea to formally test the null hypothesis that X and Y are independent.

Let ρ denote the true correlation coefficient between X and Y. The hypotheses are

$$H: \rho = 0$$

vs.

$$A: \rho \neq 0$$

To accept H is to conclude that X and Y are independent (assuming the (x_i, y_i) values in Figure 1 are a random sample from a bivariate normal distribution).

We will reject the null hypothesis only if the sample correlation coefficient, r, where

$$r = \frac{n \sum_{i=1}^{n} x_i y_i - (\sum_{i=1}^{n} x_i)(\sum_{i=1}^{n} y_i)}{\sqrt{n \sum_{i=1}^{n} x_i^2 - (\sum_{i=1}^{n} x_i)^2} \sqrt{n \sum_{i=1}^{n} y_i^2 - (\sum_{i=1}^{n} y_i)^2}}$$

is either too much less than 0 or too much greater than 0. What is too much less or too much greater is decided on the basis of

$$\frac{r}{\sqrt{\frac{1-r^2}{n-2}}}$$

When H is true this has a Student t distribution with $n - 2$ degrees for freedom. In this case, $n - 2 = 17 - 2 = 15$.

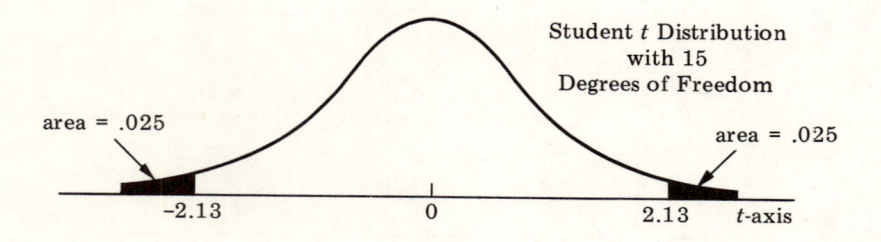

Student t Distribution with 15 Degrees of Freedom

area = .025   area = .025
−2.13   0   2.13 t-axis

If .05 is chosen to be the level of significance, H should be rejected if

$$r/\sqrt{(1-r^2)/(n-2)}$$

is either (1) less than or equal to −2.13 or 2) greater than or equal to +2.13.

Since

$$\sum_{i=1}^{17} x_i = 7,973 \qquad \sum_{i=1}^{17} x_i^2 = 4,611,291$$

$$\sum_{i=1}^{17} y_i = 8,517 \qquad \sum_{i=1}^{17} y_i^2 = 5,421,917$$

$$\sum_{i=1}^{17} x_i y_i = 4,759,470$$

it follows that

$$r = \frac{17(4,759,470) - (7,973)(8,517)}{\sqrt{17(4,611,291) - (7,973)^2} \sqrt{17(5,421,917) - (8,517)^2}}$$

$$= \frac{13,004,949}{\sqrt{14,823,218} \sqrt{19,633,300}}$$

$$= .76$$

Therefore,

$$\frac{r}{\sqrt{\frac{1 - r^2}{n - 2}}} = \frac{.76}{\sqrt{\frac{1 - (.76)^2}{17 - 2}}} = \frac{.76}{\sqrt{.02816}}$$

$$= 4.53$$

Since this number exceeds the upper critical value (2.13), our conclusion is to reject the null hypothesis.

Question 5.3.1

Having rejected H, can we conclude that excessive life-changes *cause* chronic health problems? Explain.

Question 5.3.2

Find the equation of the least squares line for these data. Plot the line on Figure 1.

REFERENCE

1. Wyler, Allen R., Ninoru Masuda, and Thomas H. Holmes, "Magnitude of Life Events and Seriousness of Illness" (*Psychosomatic Medicine*, 33, pp. 115–122).

Example 5.4 Do the feathers make the bird?

When two closely related species are crossed, the progeny will tend to have physical traits that lie somewhere "between" the corresponding traits of the parents. But what about behavioral traits? Are they similarly "mixed"? And does a hybrid whose physical traits favor one particular parent tend to have behavioral patterns that resemble those of that same parent?

One attempt at answering these questions was an experiment done with mallard and pintail ducks. A total of 11 males were studied, all were second-generation crosses. A rating scale was devised that measured the extent to which the plumage of each of the ducks resembled the plumage of the first generation's parents. A score of 0 indicated that the hybrid had the same appearance (phenotype) as a pure mallard; a score of 20 meant that the hybrid looked like a pintail. Similarly, certain behavioral traits were quantified and a second scale was constructed that ranged from 0 (completely mallard-like) to 15 (completely pintail-like).

Objective

To test whether birds that look alike act alike.

Data

Table 1
Behavioral and Plumage Indices
for 11 Mallard x Pintail Hybrids

Male	Plumage Index (x)	Behavioral Index (y)
R	7	3
S	13	10
D	14	11
F	6	5
W	14	15
K	15	15
U	4	7
O	8	10
V	7	4
J	9	9
L	14	11

Question 5.4.1

Graph the data of Table 1.

Question 5.4.2

Compute the sample correlation coefficient for these data.

Question 5.4.3

Test whether r is significantly different from 0. Use a two-sided alternative and let $P = .01$

Question 5.4.4

Is a correlation coefficient of −.50 stronger or weaker than a correlation coefficient of +.50? Explain.

REFERENCE

1. Sharpe, Roger S., and Paul A. Johnsgard, "Inheritance of Behavioral Characters in F_2 Mallard x Pintail (*Anas Platyrhynchos L.* x *Anas Acuta L.*) Hybrids" (*Behaviour*, 27, 1966, pp. 259-272).

5.3 The χ^2 test (Examples 5.5-5.9)

One of the first questions that comes up in connection with correlation data is whether X and Y are independent. If x_i and y_i are both continuous measurements, this can be answered by testing

$$H: \rho = 0$$

vs.

$$A: \rho \neq 0$$

To reject H is to conclude that X and Y are dependent.

When x_i and y_i are both qualitative the question of independence is still a relevant one but we can no longer resolve it in terms of the sample correlation coefficient. Instead, we use a procedure known as a χ^2 test.

The principles behind the χ^2 test are the same ones that were used for t tests in earlier chapters. The details, though, are quite different. In a set of qualitative, correlation data, each x_i belongs to one of r classes and each y_i, to one of c classes. (Both r and c must be greater than or equal to 2.) Thus, there are $r \times c$ different categories into which a given (x_i, y_i) might belong. The number of points in the sample that actually fall into a particular set of X- and Y-classes is called an *observed frequency* (*obs.*)

Under the null hypothesis that X and Y are independent, we can compute the number of observations that *should* have fallen into each of these classes. These numbers are called the *expected frequencies* (*exp.*)

The χ^2 test establishes the credibility of the null hypothesis by comparing the observed frequencies with the expected frequencies. If the discrepancies between the two exceed a certain limit, the null hypothesis is rejected. The quantity that is used to measure these discrepancies is

$$\sum \frac{(obs - exp)^2}{exp}$$

The sum (Σ) extends over all the $r \times c$ classes. When X and Y are, in fact, independent the behavior of $\Sigma((obs - exp)^2/exp)$ is described by a χ^2 distribution with $(r-1)(c-1)$ degrees of freedom. Examples 5.5 through 5.9 are all χ^2 tests.

Example 5.5 Agonistic behavior in mice

In Example 4.1 we looked at an experiment that tried to relate a rat's early environment to its later behavior—specifically, to its willingness to compete in a stress situation. The study described here has much the same objective but approaches it in a somewhat different way.

Two groups of mice made up the sample population; one group had been raised by their natural mothers, the other group, by "foster mice." (Each foster mouse was a female whose own litter had been removed shortly after birth.) When a mouse was three months old, it was put into a "fighting box." This was a plywood box with a partition down the middle. On the other side of the partition was a second mouse of the same age—one that had had no previous contact with the first mouse. The partition was removed and for the next six minutes the mice were carefully watched. After the six minutes were up, the experimenter recorded either a "yes" or a "no." "Yes" meant that sometime during that period the two mice began fighting; "no" meant they remained apart.

Objective

To test whether mice raised by natural mothers and mice raised by foster mothers are equally aggressive.

Data

A total of 307 mice were observed. Out of that number, 167 were raised by natural mothers, and 140 by foster mothers. Table 1 shows the group-by-group breakdown in the number fighting and the number not fighting.

Table 1
Fighting Behavior According to Type of Mother

	Natural Mother	*Foster Mother*
Number fighting	27	47
Number not fighting	140	93
	167	140

Analysis

Suppose p_{NF} denotes the true proportion of fighters among mice raised by their natural mothers. And suppose p_{FF} denotes the true proportion of fighters among mice raised by foster mothers. (From Table 1, p_{NF} is estimated by 27/167 = .164 and p_{FF}, by 47/140 = .333.) The hypotheses that we want to test are

$$H: \quad p_{NF} = p_{FF}$$

vs.

$$A: \quad p_{NF} \neq p_{FF}$$

To accept H is to conclude that "type of mother" and "willingness to fight" are independent variables.

We should accept the null hypothesis if the two sample proportions of fighters are approximately equal. Here the difference in those proportions is .333 - .164 = .169. As always, we have two possible interpretations:

1. H is true and the magnitude of the difference (that is, .169) is small enough to be attributable to chance.
2. H is false, as reflected by the size of the difference (.169).

Choosing between (1) and (2) requires a different sort of statistic than the various t tests that were used in Chapters 2, 3, and 4. The procedure that we end up with is known as a χ^2 (chi square) test.

Note, first of all, that if H were true, and there were no differences between mice raised by natural mothers and by foster mothers, a *pooled* estimate of the overall proportion of fighters would be (27 + 47)/(167 + 140) = 74/307 = .241. This means that we would "expect" 24.1 per cent of the 167 mice raised by natural mothers to fight. Therefore, while the observed frequency for that particular category was 27, the expected frequency would be (74/307) × 167 = 40.2. Similarly, the expected number of fighters among mice raised by foster mothers would be (74/307) × 140 = 33.8.

By the same reasoning, a pooled estimate of the overall proportion of mice *not* fighting is (140 + 93)/(167 + 140) = 233/307 = .759. It follows that the expected frequencies corresponding to 140 and 93 (see Table 1) are 126.8 (= (233/307) × 167) and 106.2 (= (233/307) × 140), respectively. Table 2 shows the four observed frequencies of Table 1 together with their expected frequencies under the assumption that the null hypothesis is true.

Table 2
Observed and Expected Frequencies

	Natural Mother	Foster Mother	
Number fighting	27 (40.2)	47 (33.8)	74
Number not fighting	140 (126.8)	93 (106.2)	233
	167	140	307

If the null hypothesis is true, the expected frequencies should be approximately equal to the observed frequencies. If they are not, our conclusion will be that the null hypothesis was false. To measure the magnitude of these discrepancies, we will use the quantity

$$\sum \frac{(obs - exp)^2}{exp}$$

where *obs* and *exp* refer to the observed and expected frequencies in a given category. The summation extends over all four "cells" in the table. For Table 2,

$$\sum \frac{(obs - exp)^2}{exp} = \frac{(27 - 40.2)^2}{40.2} + \frac{(47 - 33.8)^2}{33.8} + \frac{(140 - 126.8)^2}{126.8}$$

$$+ \frac{(93 - 106.2)^2}{106.2}$$

$$= 4.33 + 5.16 + 1.37 + 1.64$$

$$= 12.50$$

Of course, in order to draw any inference from this number (12.50), we need to know how the value of $\sum((obs - exp)^2/exp)$ varies from table to table (when H is true). But we do. For 2×2 contingency tables, the behavior of

$$\sum \frac{(obs - exp)^2}{exp}$$

is described by a χ^2 distribution with 1 degree of freedom.

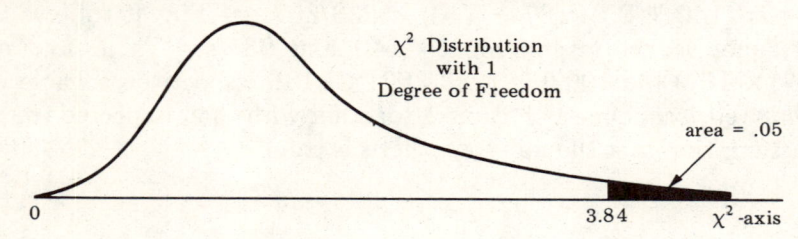

Note that as the discrepancies between observed and expected frequencies increase, so does the value of

$$\sum \frac{(\text{obs} - \text{exp})^2}{\text{exp}}$$

Furthermore, only 5 per cent of the time will a χ^2 variable with 1 degree of freedom exceed 3.84. This is another way of saying that we should reject $H: p_{NF} = p_{FF}$ in favor of $A: p_{NF} \neq p_{FF}$ at the $P = .05$ level of significance if $\sum((obs - exp)^2/exp)$ is greater than or equal to 3.84.

In this case, $\sum((obs-exp)^2/exp)$ was 12.50 so we reject the null hypothesis. Whether a mouse is raised by its own mother or by a foster mouse *does* have an effect on its aggressiveness later in life.

Question 5.5.1

Test H versus A at the $P = .01$ level of significance.

Question 5.5.2

Note that the alternative hypothesis in this example is two sided, yet the decision rule is one sided. Why?

Question 5.5.3

Chi square tests are always done by comparing observed and expected *frequencies*, as opposed to observed and expected *proportions*. Why?

REFERENCE

1. Hudgens, Gerald A., Victor H. Denenberg, and M. X. Zarrow, "Mice Reared with Rats: Effects of Preweaning and Postweaning Social Interactions upon Adult Behaviour" (*Behaviour*, 30, 1968, pp. 259-274).

Example 5.6   The sheep and goat effect

A phenomenon often studied in ESP research is the "sheep and goat" effect—the tendency for persons who believe in extrasensory perception (sheep) to perform better on ESP tests than those who do not believe (goats). In one such experiment, a group of volunteers were given a questionnaire to determine their attitudes toward ESP. The questions were all similar to the two below:

1. Have you ever had a feeling in advance that you are going to receive a particular letter on a particular day?
2. When you were participating in some game, have you ever felt compelled to bet on a certain result and won the bet?

On the basis of a subject's responses, he was classified as either (1) believing in ESP (2) not believing in ESP or (3) undecided.

All the subjects were then given a precognition test consisting of 500 blank squares. Each one was to predict which of the five standard ESP symbols ($+$, ☆ , $\bigcirc$, ≈ , $\square$) would later be entered (by a random generator) into each of the squares.

Objective

To test whether a person's precognition ability is independent of his general attitude toward ESP.

Data

Among the persons taking the test were 52 sheep and 11 goats. The scores (number of correct predictions) of these 63 subjects were classified according to whether they were above or below the average of *all* persons taking the test. The results are summarized in Table 1.

Table 1
Relative Performances by Sheep and Goats

	Sheep	Goats
Above Average	24	1
Below Average	28	10

Analysis

Let p_{SA} and p_{GA} be the true proportions of sheep and goats, respectively, who would score above average on tests of this sort. The hypotheses in question are

$$H: \quad p_{SA} = p_{GA}$$

vs.

$$A: \quad p_{SA} \neq p_{GA}$$

The analysis for testing H and A is the same as the procedure outlined in Example 5.5. There is, however, an easier way to find the expected frequencies. This shortcut is based on the fact that the expected frequencies must add up to the same row and column totals as the observed frequencies. The result is that we need to compute only one expected frequency the long way; the other three can be obtained by subtraction.

First, we calculate the expected frequency for the upper left-hand cell in Table 1. If H is true, the expected number of sheep scoring above average would be $(25/63) \times 52 = 20.6$. Since the expected frequencies in the first row must add to 25, it follows that the expected number of goats scoring above average is $25 - 20.6 = 4.4$. Likewise, the expected frequencies in the first column must add to 52, which means that the expected number of sheep scoring below average is $52 - 20.6 = 31.4$. Finally, the expected frequency in the lower right-hand cell can be calculated in two ways, as $11 - 4.4 = 6.6$ or $38 - 31.4 = 6.6$.

Table 2
Observed and Expected Frequencies

	Sheep	Goats	
Above Average	24 (20.6)	1 (4.4)	25
Below Average	28 (31.4)	10 (6.6)	38
	52	11	63

Question 5.6.1

Compute $\Sigma((obs - exp)^2/exp)$. What is the sampling distribution of $\Sigma((obs - exp)^2/exp)$ when the null hypothesis is true?

Question 5.6.2

Test H versus A at the $P = .01$ level of significance. At the $P = .05$ level of significance. What are your conclusions.

Question 5.6.3

Consider the following two contingency tables:

200	100
800	900

20	10
80	90

Note that in each case the difference in sample proportions is the same

$$\frac{200}{1000} - \frac{100}{1000} = \frac{20}{100} - \frac{10}{100}$$

Without doing any computations, explain which table would have the larger χ^2 value.

REFERENCE

1. Ryzl, Milan, "Precognition Scoring and Attitude Toward ESP" (*Journal of Parapsychology*, 32, 1968, pp. 1-8).

Example 5.7 A famous medical experiment

Until almost the end of the nineteenth century, the mortality associated with surgical operations—even minor ones—was extremely high. The major problem was infection. The germ theory as a model for disease transmission was still unknown so there was no concept of sterilization. As a result, many patients died from postoperative complications.

The breakthrough that was needed finally came when Joseph Lister, a British physician, began reading about some of the work done by Louis Pasteur. In a series of classic experiments, Pasteur had succeeded in demonstrating the part that yeasts and bacteria play in fermentation. What Lister conjectured was that human infections might have a similar organic origin. To test his theory he began using carbolic acid as an operating room disinfectant. The results were dramatic but resistance to change was so strong that it took almost ten years for his ideas to be accepted.

Objective

To test whether the mortality rate associated with amputations was independent of whether or not a disinfectant was used during the operation.

Data

Over a period of several years, before and after he had formulated his theory, Lister performed 75 amputations: 40 were done *with* carbolic acid, 35, *without*. The mortality rate for the first group was 15 per cent, as compared to 46 per cent for the second group.

Table 1
Effects of Carbolic Acid in Amputations

		Carbolic Acid Used?		
		Yes	No	
Patient Lived?	No	6	16	22
	Yes	34	19	43
		40	35	75
Per cent died		15 per cent	46 per cent	

Question 5.7.1

Define the parameters for this problem. State the null and alternative hypotheses.

Question 5.7.2

Compute the four expected frequencies under the assumption that the null hypothesis of Question 5.7.1 is true. Use the shortcut explained in Example 5.6.

Question 5.7.3

Test whether mortality rate is independent of whether or not carbolic acid is used. Use the $P = .05$ level of significance.

REFERENCE

1. Winslow, Charles, *The Conquest of Epidemic Disease* (Princeton, 1943, p. 303).

Example 5.8 Delinquency and birth order

Chi square tests for independence are not limited to the 2 × 2 format of Examples 5.5, 5.6, and 5.7. Either or both of the traits being measured can have more than two categories.

Consider the following study. The question at issue was whether a child's birth order was related to its chances of becoming a juvenile delinquent. The subjects were 1,154 girls enrolled in public high schools. They were asked to fill out a questionnaire that measured the degree to which they had exhibited delinquent behavior, in terms of criminal acts or immoral conduct. Each of the subjects was also asked to indicate her birth order, as being either (1) the oldest, (2) "in between," (3) the youngest, or (4) an only child.

Objective

To test whether delinquency and birth order are correlated.

Data

On the basis of the questionnaire, some 111 of the girls were considered to have already shown delinquent behavior. The breakdown by birth order is shown below.

Table 1
Delinquency and Birth Order

		Oldest	"In Between"	Youngest	Only Child
Delinquent?	Yes	24	29	35	23
	No	450	312	211	70
	Totals	474	341	246	93
	Per cent Yes	5.1 per cent	8.5 per cent	14.2 per cent	24.7 per cent

Analysis

The data given on the previous page are an example of a 2 × 4 contingency table. The objective is to test whether the rows are independent of the columns. That is, we want to test whether the true proportion of "yes" responses is the same from birth order to birth order. Of course, the *sample* proportions (5.1 per cent, 8.5 per cent, and so forth) are not. What we have to decide (with the help of a χ^2 test) is whether the observed variability in the sample proportions is too great to be compatible with the null hypothesis that the *true* proportions are all equal.

Let

p_{YO} = true proportion of oldest children who show delinquency tendencies

p_{YI} = true proportion of "in between" children who show delinquency tendencies

p_{YY} = true proportion of youngest children who show delinquency tendencies

p_{YOC} = true proportion of only children who show delinquency tendencies

The test for independence reduces to a choice between

$$H: \quad p_{YO} = p_{YI} = p_{YY} = p_{YOC}$$

vs.

$$A: \quad \text{not all proportions are equal}$$

As in the case of 2×2 contingency tables, we will compute expected frequencies (under the assumption that H is true) for each of the observed frequencies. If the discrepancies between the two are too large, we will reject the null hypothesis.

If H is true, the pooled estimate of the proportion of girls showing delinquent tendencies is

$$\frac{24 + 29 + 35 + 23}{474 + 341 + 246 + 93} = \frac{111}{1{,}154} = .096$$

Therefore, we would expect $(111/1{,}154) \times 474 = 45.6$ of the oldest girls to fall into this category. Likewise, we would expect 32.8 $(= (111/1{,}154) \times 341)$ and 23.7 $(= (111/1{,}154) \times 246)$ of the "in between" and youngest girls, respectively, to be delinquent. Note that having found these three expected values, the other five can be obtained by subtraction using the appropriate row and column totals. For example, the expected number of "yes" responses among only children would be $111 - 45.6 - 32.8 - 23.7 = 8.9$. Similarly, the expected number of "no" responses among oldest children would be $474 - 45.6 = 428.4$, and so on.

Table 2
Observed and Expected Frequencies

		Oldest	"In Between"	Youngest	Only Child	
Delinquent?	Yes	24 (45.6)	29 (32.8)	35 (23.7)	23 (8.9)	111
	No	450 (428.4)	312 (308.2)	211 (222.3)	70 (84.1)	1,043
		474	341	246	93	1,154

Birth Order

For any contingency table with r rows and c columns, the test statistic

$$\sum \frac{(\text{obs} - \text{exp})^2}{\text{exp}}$$

follows a χ^2 distribution with $(r-1)(c-1)$ degrees of freedom. (Does this contradict the distribution assumption made in Examples 5.5, 5.6, and 5.7?) Here $r = 2$ and $c = 4$ so $(r-1)(c-1) = (2-1)(4-1) = 3$. Note that the number that cuts off an area of .05 in the right-hand tail of the χ^2 distribution with 3 degrees of freedom is 7.81.

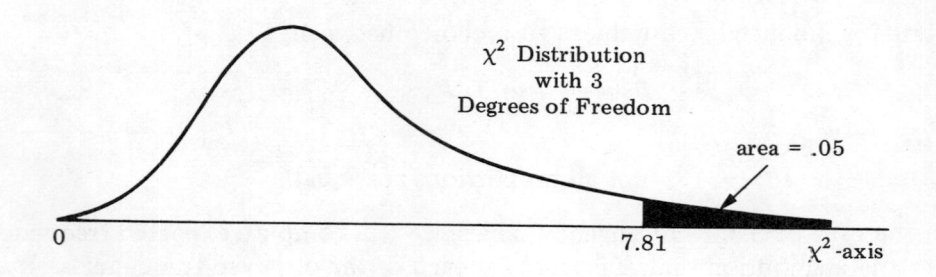

χ^2 Distribution with 3 Degrees of Freedom

area = .05

This implies that we should reject the null hypothesis at the $P = .05$ level of significance if $\Sigma((obs-exp)^2/exp)$ is greater than or equal to 7.81. But

$$\Sigma \frac{(obs-exp)^2}{exp} = \frac{(24-45.6)^2}{45.6} + \frac{(29-32.8)^2}{32.8} + \ldots + \frac{(70-84.1)^2}{84.1}$$

$$= 10.23 + .44 + \ldots + 2.36$$

$$= 42.5$$

Our conclusion, then, is to reject the null hypothesis: Birth order does have an effect on a girl's chances of becoming a delinquent.

Question 5.8.1

Why would it be incorrect to write the alternative hypothesis as

$$A: p_{YO} \neq p_{YI} \neq p_{YY} \neq p_{YOC} ?$$

Question 5.8.2

In computing the expected frequencies, we found three directly and the remaining five by subtraction. Does it matter *which* three are found directly? Explain. What interpretation can you give to the phrase "degree of freedom" in this problem?

Question 5.8.3

Suppose this same sort of study was done in a lower-class neighborhood and also in an upper-class neighborhood. Do you think the same trends would hold for both populations?

REFERENCE

1. Nye, Francis Iven, *Family Relationships and Delinquent Behavior* (Wiley, 1958, p. 37).

Example 5.9 The psychology of small cars

How is the American car buyer reacting psychologically to the trend toward small cars? Does he view these cars as being something fundamentally different or simply as scaled-down versions of models from years past? And what, if anything, do his attitudes toward small cars reveal about his own personality?

Recently a market research team did a survey to try to answer some of these questions. A total of 250 persons were contacted. All were adults living in a large metropolitan area. They were first asked to fill out a 16-item self-perception questionnaire. On the basis of their responses they were put into one of three distinct personality types: (1) cautious conservative, (2) middle-of-the-roader, and (3) confident explorer. They were also asked to give their overall opinion of small cars. Here there were three possible responses: (1) favorable, (2) neutral, and (3) unfavorable.

Before the data were tabulated it was hypothesized that there *would* be attitude differences from personality type to personality type. Specifically, it was felt that the subjects who tended to accept change more readily (that is, the "confident explorers") would react the most favorably to small cars. As it turned out, the exact opposite was true.

Objective

To see whether there is any correlation between personality type and a person's attitude toward small cars.

Data

Table 1
Self-perception and Attitude Toward Small Cars

		Cautious conservative	Self-perception Middle-of-the-roader	Confident explorer
Opinion of Small Cars	Favorable	79	58	49
	Neutral	10	8	9
	Unfavorable	10	34	42

Question 5.9.1

State the parameters and hypotheses relevant to this problem.

Question 5.9.2

Compute the expected values for the data in Table 1.

Question 5.9.3

Test the hypotheses of Question 5.9.1. Use the $P = .01$ level of significance. What conclusions would you draw about personality type and attitude toward small cars?

REFERENCE

1. Jacobson, Eugene, and Jerome Kossoff, "Self-percept and Consumer Attitudes Toward Small Cars" (in *Consumer Behavior in Theory and in Action*, Steuart Henderson Britt, ed., Wiley, 1970, pp. 126–129).

Mathematicians are like Frenchmen;
Whatever you say to them,
They translate into their own language
And forthwith it is something entirely different.

Goethe

Chapter 6

Nonparametric Methods

6.1 Introduction

It is sometimes easy to forget that the inference procedures in Chapters 2, 3, 4, and 5 are not always valid. Having observed $x_1, x_2, \ldots, x_n$, we cannot immediately conclude, for example, that

$$\frac{\overline{X} - \mu_0}{s/\sqrt{n}}$$

follows a Student t distribution with $n-1$ degrees of freedom. Only if the x_i's come from a population that is normal (and has mean μ_0) will that statement be true. The problem is that if $(\overline{X} - \mu_0)(s/\sqrt{n})$ does not follow a Student t distribution, the t test, itself, becomes suspect.

This raises an obvious question: What happens if the population being sampled is not normal but we do a t test anyway? Outwardly, nothing—but, "internally," the basic properties of our inference procedure may be drastically altered. Specifically, the *true* level of significance—that is, the probability of committing a Type I error—may be something quite different than the *nominal* level of significance, P.

For example, suppose we are doing a two-sided, one-sample t test at the $P = .05$ level of significance with $n = 20$ observations. And suppose the assumption of normaility *is* true. Then

$$P\left(\frac{\overline{X} - \mu_0}{s/\sqrt{20}} \geq 2.09\right) = .025 = P\left(\frac{\overline{X} - \mu_0}{s/\sqrt{20}} \leq -2.09\right)$$

That is, for $P = .05$, the numbers -2.09 and $+2.09$ define the critical regions for testing $H: \mu = \mu_0$ versus $A: \mu \neq \mu_0$.

Now, suppose the assumption of normality is *not* true. Then

$$P\left(\frac{\overline{X} - \mu_0}{s/\sqrt{20}} \geq 2.09\right)$$

will not equal .025; neither will

$$P\left(\frac{\overline{X} - \mu_0}{s/\sqrt{20}} \leq -2.09\right)$$

They may be close—say, .024 or .026—but they may be very different, maybe .350 or .001. The implication is that if we were to do a t test on these data, and use -2.09 and $+2.09$ as critical values, the true level of significance would not be .05.

Of course, the magnitude of the discrepancy between the true level of significance and the nominal level of significance will depend on how badly the assumption of normality is violated. A population that is just slightly skewed will not affect the distribution of $(\overline{X} - \mu_0)/(s/\sqrt{n})$ as much as one that is U-shaped. Unfortunately, since we seldom know the exact form of the population being sampled, there is no way to accurately predict how large the error is going to be.

Actually, the situation is not as grim as it might seem: The behavior of $(\overline{X} - \mu_0)/(s/\sqrt{n})$ is not totally unpredictable. Quite the contrary. Mathematical statisticians have shown that the t statistic is remarkably *robust* with respect to departures from normality. That is, unless $x_1, x_2, \ldots, x_{20}$ come from a very bizarre distribution,

$$P\left(\frac{\overline{X} - \mu_0}{s/\sqrt{20}} \geq 2.09\right) \quad \text{and} \quad P\left(\frac{\overline{X} - \mu_0}{s/\sqrt{20}} \leq -2.09\right)$$

will be much closer to .025 than to .350 or .001.

Nevertheless, in response to the uncertainty surrounding the distribution of $(\overline{X} - \mu_0)/(s/\sqrt{n})$ for arbitrary $x_1, x_2, \ldots, x_n$, a set of inference procedures have been developed that require far weaker assumptions than the t test. These are known as *nonparametric statistics*. Typically, their only stipulation is that the x_i's are a random sample from a *continuous* distribution. The precise shape of that distribution—whether it be normal, rectangular, skewed, U-shaped, or anything else—has no effect on the probability of committing a Type I error.

In this chapter we will examine some of the more standard nonparametric techniques. The examples themselves range from a quantitative study of international politics to an experiment comparing the preening behavior of male and female fruit flies. In each instance, try to recognize what there was about the data that suggested that a nonparametric analysis would be appropriate.

6.2 Rank tests

Non-parametric methods come in a bewildering assortment of formats and types. Corresponding to every parametric procedure we have seen there are any number of nonparametric analogs, each based on slightly different assumptions or testing a slightly different hypothesis. Some are so easy to use they can be done without even pencil and paper; others require a computer.

There is no way to characterize all of these procedures. Many of them, though, require that the original observations be replaced by ranks. For example, suppose we do a two-sample problem and record the following set of x's and y's:

Treatment 1	Treatment 2
$x_1 = 3.6$	$y_1 = 4.4$
$x_2 = 7.5$	$y_2 = 4.5$
$x_3 = 6.0$	$y_3 = 5.8$
$x_4 = 5.2$	

Arranging these from smallest to largest gives

$$3.6 < 4.4 < 4.5 < 5.2 < 5.8 < 6.0 < 7.5$$

We define the *rank* of an observation as its position in the combined sample. Here, 3.6 would be assigned rank 1, 4.4, rank 2, and so on.

Treatment 1 Ranks	Treatment 2 Ranks
1	2
7	3
6	5
4	

Clearly, the way the ranks are distributed between the two treatments can be used to test for shifts in location in much the same way the original x_i's and y_i's could. If the median ($\tilde{\mu}_1$) for Treatment 1 was greater than the median ($\tilde{\mu}_2$) for Treatment 2, the ranks in the Treatment 1 sample would tend to exceed the ranks in the Treatment 2 sample. If the discrepancy was large enough, we would be justified in rejecting H: $\tilde{\mu}_1 = \tilde{\mu}_2$ in favor of A: $\tilde{\mu}_1 > \tilde{\mu}_2$.

The solutions to Examples 6.1, 6.2, 6.6, and 6.7 are all based on ranks. The first two are known as *Wilcoxon rank sum tests*. These are analogs of the two-sample t test. The last two are somewhat more general procedures known as *Kruskal-Wallis tests*. They can compare the locations of k populations, where k is any integer greater than 2.

There is also a nonparametric version of the correlation coefficient. Instead of relating the x_i's to the y_i's, it associates the *ranks* of the x_i's with the *ranks* of the y_i's. This is done using a statistic known as the *Spearman rank correlation coefficient*. Its advantage over the more familiar correlation coefficient of Chapter 5 is that it remains valid even if the (x_i, y_i)'s are not a random sample from a bivariate normal. Examples 6.4, 6.5, describe applications of the rank correlation coefficient.

There is a rather extensive set of nonparametric procedures that do not involve ranks. These are called *runs tests*. They are often used to determine whether the fluctuations in an ordered series of observations are random. In Example 6.3, we use a runs test to examine the relationship between labor and management at a time when unions were still in their infancy.

Example 6.1 The politics of war

Spiraling paranoia is a national state of mind that always precedes a declaration of war. Country B thinks that Country A has made a hostile gesture. They respond. Then Country A responds to Country B's response, and the spiral has begun. World War I is an example. On June 28, 1914, a Serbian nationalist assassinated the Archduke Francis Ferdinand, heir-apparent to the throne of Austria-Hungary. The aftermath was a rapid-fire sequence of military posturings, political innuendoes, charges and countercharges that escalated a Balkan dispute into a world war.

Historians have tried to become more objective about what happened in that summer of 1914 by quantifying, on an almost day-by-day basis, the hostilities directed toward each of the countries involved. A scale was devised that assigned a hostility intensity level (S_i) to the actions taken by the enemy during the ith time period. These actions ranged from inflammatory statements made by diplomats to overt military operations. The same scale was used to measure the hostility intensity (R_i) of the actions taken in retaliation.

In this example we compare the differences, $S_i - R_i$, for the Dual Alliance (Austria-Hungary, Germany) and the Triple Entente (England, France, Russia). A negative value for $S_i - R_i$ indicates that a country overreacted to the threats of its enemies; a positive value means it underreacted. Since there is no *a priori* reason why the distribution of $S_i - R_i$ values for either alliance should be normal, a nonparametric analysis is appropriate.

Objective

To compare the intensities of the political and military reactions of the Dual Alliance and the Triple Entente to the events immediately preceding the beginning of World War I.

Data

Table 1 shows the input hostility intensities (S_i) and the output hostility intensities (R_i) recorded for the Triple Entente. The time period covered extends from the week of the assassination through August 4. Table 2 gives the same information for the Dual Alliance.

Table 1
Average Intensity Levels of Violence for the Triple Entente

Time period	By Enemy (Dual Alliance), S_i	In response, R_i
June 27–July 2	4.25	4.38
July 3–16	4.25	2.58
July 17–20	3.00	2.62
July 21–25	2.83	4.28
July 26	5.38	3.68
July 27	5.37	4.95
July 28	5.87	4.68
July 29	6.06	5.07
July 30	4.64	4.60
July 31	5.10	5.50
August 1–2	6.30	5.90
August 3–4	5.88	6.03

Analysis

What concerns us here are not the increases in the S_i's or the R_i's but, rather, the comparison of $S_i - R_i$ for the two coalitions. Did both alliances react to events with equal restraint (or lack of restraint) or was there a tendency for one side to underreact and the other to overreact? Table 3 lists the two sets of intensity differences.

Table 2
Average Intensity Levels of Violence for the Dual Alliance

Time period	By Enemy (Triple Entente), S_i	In response, R_i
June 27–July 2	4.38	4.25
July 3–16	4.38	3.00
July 17–20	2.58	2.83
July 21–25	2.62	5.38
July 26	4.28	5.37
July 27	3.68	5.87
July 28	4.95	6.06
July 29	4.68	4.64
July 30	5.07	5.10
July 31	4.60	6.30
August 1–2	5.50	5.88
August 3–4	5.90	6.08

Table 3
Difference in Intensity Between Input Behavior (S_i) and Output Behavior (R_i)

Time period	Triple Entente, $S_i - R_i$	Dual Alliance, $S_i - R_i$
June 27–July 2	−0.13	+0.13
July 3–16	+1.67	+1.38
July 17–20	+0.38	−0.25
July 21–25	−1.45	−2.76
July 26	+1.70	−1.09
July 27	+0.42	−2.19
July 28	+1.19	−1.11
July 29	+0.99	+0.04
July 30	+0.04	−0.03
July 31	−0.40	−1.70
August 1–2	+0.40	−0.38
August 3–4	−0.15	−0.18

Not knowing the distribution of the $S_i - R_i$ values, we will replace each one with its rank and do a Wilcoxon rank sum test on the results. If the 24 differences were arranged from largest (most positive) to smallest (most negative), the July 26 difference for the Triple Entente (+1.70) would be assigned a rank of 1; the July 3–16 difference for the Triple Entente (+1.67) would have rank 2, and so on. The lowest rank (24) would go to the July 21–25 difference for the Dual Alliance (−2.76). Table 4 shows the entire set of ranks.

Table 4
Ranks for $S_i - R_i$ Differences

Time period	Triple Entente	Dual Alliance
June 27–July 2	13	9
July 3–16	2	3
July 17–20	8	16
July 21–25	21	24
July 26	1	19
July 27	6	23
July 28	4	20
July 29	5	10.5
July 30	10.5	12
July 31	18	22
August 1–2	7	17
August 3–4	14	15

Suppose P_D and P_T denote the two population distributions of hostility differences for the Dual Alliance and the Triple Entente, respectively. The hypotheses that we want to test are

$$H: P_D \text{ and } P_T \text{ are identical}$$

vs.

$$A: P_D \text{ and } P_T \text{ differ with respect to location}$$

One way of deciding between H and A is to base a decision rule on W, the sum of the ranks for the Dual Alliance. Since there are 24 ranks in the combined sample, the total rank sum is

$$\sum_{i=1}^{24} i = 1 + 2 + \ldots + 24 = 300$$

It follows that if H were true, the sum of the Dual Alliance ranks would be somewhere near half that number. If P_D were shifted to the "left" of P_T, W would be considerably *larger* than 150; if it were shifted to the right, it would be considerably smaller. (Why?)

From Table 4,

$$W = 9 + 3 + 16 + 24 + 19 + 23 + 20 + 10.5 + 12 + 22 + 17 + 15$$
$$= 190.5$$

To complete the analysis we need to know the sampling distribution of W, so that the observed value (190.5) can be properly interpreted. For sample sizes less than or equal to 10, the exact distribution of W has been tabulated (see, for example, Hollender and Wolfe, 1973). For larger sample sizes, a normal approximation is used. If n and m are the two sample sizes—and if W is the rank sum for the sample of size n—the distribution of

$$W^* = \frac{W - \frac{n(m+n+1)}{2}}{\sqrt{\frac{mn(m+n+1)}{12}}}$$

is approximated by the standard normal. Therefore, if H and A are to be tested at the $P = .05$ level of significance, H is rejected if W^* is either (1) less than or equal to -1.96 or (2) greater than or equal to $+1.96$. In this case, $n = 12$ and $m = 12$ so that

$$W^* = \frac{190.5 - \frac{12(25)}{2}}{\sqrt{\frac{144(25)}{12}}} = \frac{40.5}{17.32}$$

$$= 2.34$$

Question 6.1

What conclusion is reached and how would it be interpreted?

Question 6.2

Test H versus A at the $P = .01$ level of significance.

Question 6.3

What is the relationship between $n(m + n + 1)/2$ and W? Between $((mn(m + n + 1))/12)^{1/2}$ and W?

REFERENCES

1. Hollander, Myles, and Douglas A. Wolfe, *Nonparametric Statistical Methods* (Wiley, 1973, pp. 272-282).
2. Holsti, Ole R., Robert C. North, and Richard A. Brody, "Perception and Action in the 1914 Crisis" in *Quantitative International Politics*, J. David Singer, ed. (The Free Press, 1968, pp. 123-158).

Example 6.2 Preening behavior in *Drosophila melanogaster*

In most two-sample problems there is no way to know whether the populations being sampled are normal. We can draw histograms of the data to see how bell shaped their distributions are, but typically the two sample sizes are so small that this is not very enlightening. Sometimes, though, the nonnormality is obvious. If one or two observations are either much smaller or much larger than all the others, we can safely assume that the populations being sampled are *not* normal. When this is the case a nonparametric procedure is warranted.

This was precisely the reason for using a rank sum test on the results described in this example. The data consist of 30 observations, almost all less than 3.0. But two numbers exceed 10.0. If these were included with all the others, as a *t* test would do, they would have a disproportionately large influence on the final results. By replacing the original set of measurements with their ranks we prevent this from happening.

Objective

To contrast the average lengths of time spent preening by male fruit flies and female fruit flies.

Procedure

The experiment was done using a small chamber that was divided into two parts by a glass partition. The fruit flies (*Drosophila melanogaster*) of one sex were put into one side of the chamber and a single fly of the same sex into the other. All flies were the same sex so that courtship rituals would not be a factor.

For three minutes the preening behavior of the single fly was carefully watched. One of the variables observed was the average length of time it spent preening (see Table 1). Eventually the entire experiment was done 30 times, 15 times using male flies and 15 times using female flies.

Data

Table 1
Average Preening Time per Bout (sec)

Male	*Female*
2.3	3.7
1.9	5.4
3.3	2.2
2.9	11.7
2.2	2.8
1.3	2.4
2.2	4.0
2.4	2.8
2.1	2.0
1.2	2.8
2.0	2.4
2.7	2.4
2.3	2.9
1.9	10.7
1.2	3.2

Question 6.2.1

Rank the combined sample

Average Preening Time per Bout (sec)

Male	Rank	Female	Rank
2.3		3.7	
1.9		5.4	
3.3		2.2	
2.9		11.7	
2.2		2.8	
1.3		2.4	
2.2		4.0	
2.4		2.8	
2.1		2.0	
1.2		2.8	
2.0		2.4	
2.7		2.4	
2.3		2.9	
1.9		10.7	
1.2		3.2	

Question 6.2.2

Compute the rank sum (W) for the male fruit flies.

Question 6.2.3

State the hypotheses that these data would be used to test.

Question 6.2.4

Using the normal approximation to the distribution of W, test the hypotheses of Question 6.2.3. Use the $P = .05$ level of significance. Should the alternative be one sided or two sided?

REFERENCE

1. Connolly, Kevin, "The Social Facilitation of Preening Behaviour in *Drosophila Melanogaster*" (*Animal Behavior*, **16**, pp. 385–391).

Example 6.3 U.S. labor disputes

The first widespread labor dispute in the United States occurred in 1877. Railroads were the target and workers were idled from Pittsburgh to San Francisco. That first dispute may have been a long time coming but workers were quick to recognize what a powerful weapon the strike really was—between 1881 and 1905 they called 36, 757 more.

In this example we look at those 36,757 strikes year by year. Specifically, we will focus on the proportion each year that were considered successful. (A successful strike, by definition, was one where most or all of the workers' demands were met.) An obvious question is whether these proportions exhibit any patterns or trends. We might expect them to gradually increase as unions acquired more and more power. Or, we might think that years of high success rates would tend to alternate with years of low success rates, indicating a give-and-take relationship on the part of labor and management. Still another theory would be that the numbers show *no* patterns and behave like a random sequence.

A standard way of investigating the reandomness of a sequence of numbers is with a nonparametric procedure known as a *runs test*. There are actually many such tests. Here we will consider runs *above and below the median*.

Objective

To test for nonrandomness in the yearly proportions of successful U.S. labor strikes during the period from 1881 to 1905.

Data

Table 1
Strike Statistics (1881–1905)

Year	Number of strikes	Per cent successful
1881	451	61
1882	454	53
1883	478	58
1884	443	51
1885	645	52
1886	1,432	34
1887	1,436	45
1888	906	52
1889	1,075	46
1890	1,833	52
1891	1,717	37
1892	1,298	39
1893	1,305	50
1894	1,349	38
1895	1,215	55
1896	1,026	59
1897	1,078	57
1898	1,056	64
1899	1,797	73
1900	1,779	46
1901	2,924	48
1902	3,161	47
1903	3,494	40
1904	2,307	35
1905	2,077	40

Analysis

Figure 1 is a plot of the proportions of successful strikes for the 25 years in question.

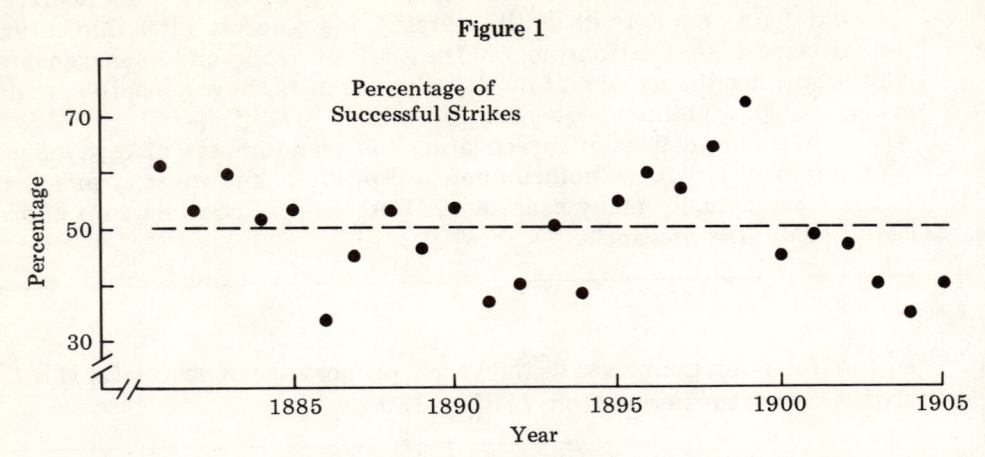

Figure 1

Percentage of Successful Strikes

One way to examine the randomness of a series of ordered numbers is to look at the way they fluctuate above and below the sample median. For these data the median is the 13th largest observation, 50. This is shown as a horizontal line in Figure 1.

We define a *run* as an uninterrupted series of observations on one side of the median. Starting with the figure for 1881, there is a run of length 5 *above* the median (61, 53, 58, 51, 52). This is followed by a run of length 2 *below* the median (34, 45), a run of length 1 *above* the median (52), and so on. Altogether, there are $R = 8$ runs in these 25 observations.

Is a total of 8 runs compatible with the null hypothesis that the observations are random with respect to time? Note that if there were a very sharp trend in the data, there would be only 2 runs; on the other hand, if the observations alternated above and below the median there would be 24. It follows that if the actual number of runs is too close to either of these extremes, we will conclude that the numbers are nonrandom.

If n is the number of observations above (and below) the median, it can be shown that when the points are random the expected number of runs is equal to

$$\mu = n + 1$$

The standard deviation is

$$\sigma = \sqrt{\frac{n(n-1)}{2n-1}}$$

Furthermore, the distribution of

$$R^* = \frac{R - \mu}{\sigma}$$

is approximated by a standard normal. This means that we should reject H at the $P = .05$ level of significance if R^* is either (1) less than or equal to -1.96 or (2) greater than or equal to 1.96.

Question 6.3.1

Compute μ and σ for these data and carry out the test.

Question 6.3.2

Is it possible for this test to accept the null hypothesis at a very high level of significance and yet for the data to be obviously nonrandom? Explain.

Question 6.3.3

What other properties of runs might be useful in testing a set of data for randomness?

REFERENCE

1. Craf, John R., *Economic Development of the U.S.* (McGraw-Hill, 1952, pp. 368–371).

Example 6.4 Suicide rates

Ever since Émile Durkheim published *Suicide: A Study in Sociology* in 1897, the phenomenon of suicide has been widely studied as a means of evaluating the levels of stress prevailing in a society. Over the years, suicide rates have been shown to follow some very definite patterns. They are much higher for males than for females and for whites than for blacks. They are high for the elderly, the unmarried, and the childless. They are low in times of war and high in times of peace.

To a large extent, what causes these patterns is still unknown. Alienation, though, is undoubtedly one of the factors high on the list. This might explain why suicide rates are higher in urban areas than in smaller communities. It also suggests that cities whose populations are more transient should have higher suicide rates than cities whose populations are more settled.

Objective

To test whether there is a significant correlation in large urban areas between suicide rate and population mobility.

Data

A total of 25 large American cities were included in the sample. Table 1 shows the suicide rate and mobility index for each one. Lower values for the mobility index correspond to populations that are more transient.

Table 1
Suicide Rate and Mobility Index for 25 American Cities

City	Mobility index	Suicides per 100,000
New York	54.3	19.3
Chicago	51.5	17.0
Philadelphia	64.6	17.5
Detroit	42.5	16.5
Los Angeles	20.3	23.8
Cleveland	52.2	20.1
St. Louis	62.4	24.8
Baltimore	72.0	18.0
Boston	59.4	14.8
Pittsburgh	70.0	14.9
San Francisco	43.8	40.0
Milwaukee	66.2	19.3
Buffalo	67.6	13.8
Washington	37.1	22.5
Minneapolis	56.3	23.8
New Orleans	82.9	17.2
Cincinnati	62.2	23.9
Newark	51.9	21.4
Kansas City	49.4	24.5
Seattle	30.7	31.7
Indianapolis	66.1	21.0
Rochester	68.0	17.2
Jersey City	56.5	10.1
Louisville	78.7	16.6
Portland	33.2	29.3

Analysis

Our objective is to determine whether mobility index (X) and suicide rate (Y) are correlated. That is, does knowing a city's mobility index help us predict its suicide rate?

If we could assume that the (x_i, y_i) pairs were a random sample from a bivariate normal distribution, the appropriate analysis would be to compute the sample correlation coefficient, r. This was the procedure followed in Examples 5.3 and 5.4. Once r is computed, we can test $H: \rho = 0$ against $A: \rho \neq 0$, where ρ is the *true* correlation coefficient between X and Y. If H is rejected, the conclusion is that X and Y are correlated.

If for some reason the normality assumption seems unreasonable, the analysis proceeds a little differently. Instead of computing r we calculate r_s, the Spearman rank correlation coefficient. The difference between the two is that r measures the correlation between x_i and y_i whereas r_s measures the correlation between the rank of x_i and the rank of y_i. And, more importantly, the validity of r_s is not dependent on what kind of distribution the original observations came from.

To compute r_s for these data we first need to assign ranks to both the suicide rates and the mobility indices. First, consider the suicide rates. If they were arranged from largest to smallest, San Francisco (suicide rate = 40.0/100,000) would receive a rank of 1; Seattle (suicide rate = 31.7/100,000), a rank of 2, and so on. Jersey City has the lowest suicide rate and would be assigned a rank of 25. When two cities have exactly the same suicide rate they share the average of the ranks that they would have been assigned had their rates been slightly different. Los Angeles and Minneapolis, for example, both have rates of 23.8. Had there been a slight difference between the two, one would have been ranked 7, the other, 8. Therefore, each is given the rank 7.5 ($=\frac{7+8}{2}$). The entire set of suicide ranks is shown in Column 3 of Table 2.

The mobility indices can be treated in the same way. Recall that low values of the mobility index correspond to high degrees of mobility. To be consistent with the way suicide rates were ranked, the mobility indices will be ranked from smallest to largest. That is, Los Angeles (mobility index = 20.3) is ranked 1 and New Orleans (mobility index = 82.9), 25 (see Column 2 of Table 2).

The Spearman rank correlation coefficient is defined by

$$r_s = 1 - \frac{6 \sum_{i=1}^{n} d_i^2}{n(n^2 - 1)}$$

where n is the number of subjects (in this case, 25). Like r, values for r_s range between -1 and $+1$, inclusive. Numbers close to -1 indicate a strong *negative* correlation, numbers close to $+1$, a strong *positive* correlation, and numbers close to 0, *no* correlation.

To test whether the observed r_s is significantly different from 0, we use the fact that

$$\frac{r_s}{1/\sqrt{n-1}} = \sqrt{n-1}\ r_s$$

has approximately a standard normal distribution when X and Y are independent. It follows that the null hypothesis of independence should be rejected at, say, the $P = .05$ level of significance if $\sqrt{n-1}\ r_s$ is either (1) ≤ -1.96 or (2) $\geq +1.96$. (Why?)

Table 2
Ranks for Mobility Indices and Suicide Rates

City	Mobility Rank (x_i)	Suicide Rank (y_i)	$d_i = y_i - x_i$	d_i^2
New York	11	13.5	2.5	6.25
Chicago	8	19	11	121.00
Philadelphia	17	16	-1	1.00
Detroit	5	21	16	256.00
Los Angeles	1	7.5	6.5	42.25
Cleveland	10	12	2	4.00
St. Louis	16	4	-12	144.00
Baltimore	23	15	-8	64.00
Boston	14	23	9	81.00
Pittsburgh	22	22	0	0
San Francisco	6	1	-5	25.00
Milwaukee	19	13.5	-5.5	30.25
Buffalo	20	24	4	16.00
Washington	4	9	5	25.00
Minneapolis	12	7.5	-4.5	20.25
New Orleans	25	17.5	-7.5	56.25
Cincinnati	15	6	-9	81.00
Newark	9	10	1	1.00
Kansas City	7	5	-2	4.00
Seattle	2	2	0	0
Indianapolis	18	11	-7	49.00
Rochester	21	17.5	-3.5	12.25
Jersey City	13	25	12	144.00
Louisville	24	20	-4	16.00
Portland	3	3	0	0
				1,199.50

For these data

$$r_s = 1 - \frac{6(1,199.5)}{25(625-1)} = 1 - .46$$

$$= .54$$

Therefore,

$$\sqrt{n-1}\ r_s = \sqrt{25-1}\ (.54) = 2.64$$

Since 2.64 exceeds 1.96, our conclusion, at the $P = .05$ level of significance, is that suicide rate and mobility index *are* correlated. More specifically, cities whose populations are more transient tend to have higher suicide rates.

Question 6.4.1

For this same sample size, what is the largest value of r_s that will lead us to accept the null hypothesis of no correlation?

Question 6.4.2

Is the observed correlation coefficient significant at the $P = .01$ level of significance?

Question 6.4.3

Can we conclude from this analysis that greater mobility is a contributory cause to high suicide rates? Explain.

Question 6.4.4

Suppose the mobility indices had been ranked from largest to smallest instead of from smallest to largest. How would the numerical value of r_s have changed?

REFERENCE

1. Young, Pauline V., and Calvin Schmid, *Scientific Social Surveys and Research* (Prentice-Hall, 1966, p. 319).

Example 6.5 Adolescent problems

The need to use the rank correlation coefficient, r_s, instead of the usual sample correlation coefficient, r, can arise in two different contexts: (1) the data are numerical but we elect not to assume that they represent a random sample from a bivariate normal distribution, or (2) the data are ranks to begin with. The suicide-mobility data of the previous example typified the first situation; here we look at the second.

The subjects in this study were some 7,000 12th grade students from 57 different high schools throughout Illinois. Each student was given the Mooney Problem Check List. This is a questionnaire consisting of 330 problems that often trouble adolescents. School officials hoped that the results would pinpoint the most serious problem areas and, in so doing, revitalize their high school counseling program.

> **Objective**
>
> To determine whether counseling programs for adolescent boys should focus on the same problems as counseling programs for adolescent girls.

Data

The 330 individual problems on the Mooney questionnaire were grouped into 11 general categories. Table 1 gives the rank order frequencies with which these categories were mentioned by the 7,000 respondents. That is, the most frequently cited problem area among the boys was "adjustment to school work"; the second most frequently cited was "future: vocational and educational," and so on.

Table 1
Patterns in Adolescent Problems for 12th Grade Males and Females

Problem Area	Rank for Males	Rank for Females
Adjustment to school work	1	2
Curriculum and teaching procedures	3	4
Future: vocational and educational	2	5
Personal-psychological relations	8	1
Social-psychological relations	6	3
Social and recreational activities	4.5	6
Courtship, sex, marriage	4.5	8
Health and physical development	8	7
Finances, living conditions, and employment	8	9.5
Home and family	11	9.5
Morals and religion	10	11

Question 6.5.1

Define an appropriate null hypothesis and alternative hypothesis for these data.

Question 6.5.2

Compute r_s.

Question 6.5.3

Test whether r_s is significantly different from 0.

Question 6.5.4

If the x_i's and y_i's do not come from a bivariate normal distribution it is not valid to test $H: \rho = 0$ on the basis of r. On the other hand, if the data *are* a random sample from a bivariate normal, we can still compute r_s and test $H: \rho = 0$ accordingly. Why should we ever *not* use r_s? What is gained by using r when the assumptions it requires are met?

REFERENCE

1. Blair, Glenn Myers, Stewart R. Jones, and Ray H. Simpson, *Educational Psychology* 4th ed., (Macmillan, 1975, pp. 462–463).

Example 6.6 Factors affecting well productivity

In the experiments we have seen up to this point the number of treatments to be compared was at most two. Situations arise, though, where three or four or, in general, k treatments need to be compared. These are known generically as *k-sample problems*. As was the case with the one- and the two-sample problems, the k-sample problem has both parametric and nonparametric solutions, the appropriate choice depending on the assumptions that can be made about the k populations. In the next two examples we will look at the standard nonparametric solution, a rank procedure known as the *Kruskal-Wallis test*.

Geologists are always concerned with the reasons why one well is better than another in terms of water quality, productivity, longevity, and so on. Are these well-to-well differences random or are they the product of specific geological factors? The question is obviously a practical one. If the differences do have assignable causes, future wells should be restricted to areas that are geologically favorable.

Objective

To test whether the type of rock in which a water well is located affects its yield.

Data

A total of 80 water wells in Central Pennsylvania were studied. They were all dug in one of five types of rock, upper sandy dolomite, nittany dolomite, bellefonte dolomite, limestone, and shale. The productivities for the 80 wells were ranked from lowest (1) to highest (80).

Analysis

The null hypothesis for these data is that well productivity is unaffected by rock type. If that were true, we would expect the average of the ranks for the five sets of wells to be fairly similar. If it were not true, at least one of the groups would have ranks that tended to be relatively high or relatively low.

To choose between these two alternatives, Kruskal and Wallis proposed the statistic

$$H = \frac{12}{N(N+1)} \sum_{j=1}^{k} \frac{R_j^2}{n_j} - 3(N+1)$$

where

N = total sample size

R_j = rank sum for the jth treatment group

n_j = sample size for the jth treatment group

Table 1
Productivity Ranks for Wells in Five Rock Types

Upper sandy dolomite	Nittany dolomite	Bellefonte dolomite	Limestone	Shale
9	12	1	5	8
18	24	2	13	10
28	29	3	21	14
41	33	4	22	19
43	39	6	27	23
44	46	7	31	25
45	49	11	42	26
50	53	15	54	35
58	57	16	56	47
59	67	17	66	55
60	72	20	69	
61	77	30		
62	78	32		
63	79	34		
65	80	36		
68		37		
70		38		
71		40		
73		48		
74		51		
75		52		
76		64		

For these data,

$$R_1 = 9 + 18 + 28 + \ldots + 76 = 1{,}213$$

$$R_2 = 12 + 24 + 29 + \ldots + 80 = 795$$

$$R_3 = 1 + 2 + 3 + \ldots + 64 = 564$$

$$R_4 = 5 + 13 + 21 + \ldots + 69 = 406$$

$$R_5 = 8 + 10 + 14 + \ldots + 55 = 262$$

Also, $n_1 = 22$, $n_2 = 15$, $n_3 = 22$, $n_4 = 11$, $n_5 = 10$, $N = 80$, and $k = 5$. Therefore,

$$H = \frac{12}{80(80+1)} \left\{ \frac{(1{,}213)^2}{222} + \frac{(795)^2}{15} + \frac{(564)^2}{22} + \frac{(406)^2}{11} + \frac{(262)^2}{10} \right\} - 3(80+1)$$

$$= (.001852) \left\{ 145{,}323.81 \right\} - 243 = 269.14 - 243$$

$$= 26.14$$

To complete the analysis, and put the number 26.14 in proper perspective, we need to know the sampling distribution of H. Kruskal and Wallis were able to show that when the null hypothesis is true, the behavior of H can be approximated by a χ^2 distribution with $k-1$ degrees of freedom. Furthermore, the values of H least compatible with the null hypothesis are the ones in the right-hand tail of that distribution. In this case, with k equal to 5, we would reject the null hypothesis at the $P = .05$ level of significance if H is greater than or equal to 9.49.

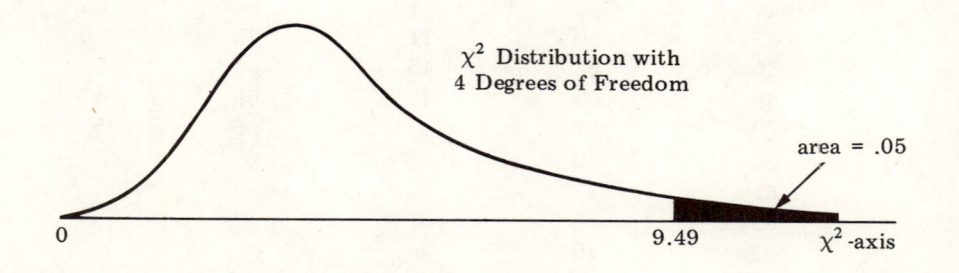

Since H was computed to be 26.14, we reject the null hypothesis and conclude that rock type *is* a factor in determining water well productivity.

Question 6.6.1

Test these same hypotheses at the $P = .01$ level of significance.

Question 6.6.2

Suppose there were 16 wells in each of the five rock types. What would be the expected average rank for each of the groups if the null hypothesis were true?

Question 6.6.3

In a general k-sample problem, will the H statistic be an effective way of testing whether all the populations have the same dispersion? Explain.

REFERENCE

1. Siddiqui, S. H., and R. R. Parizek, "Application of Nonparametric Statistical Tests in Hydrogeology" (*Ground Water*, 10, 1972, pp. 26-31).

Example 6.7 The 1969 draft lottery

On December 1, 1969, a lottery was held in Selective Service headquarters in Washington, D.C. to determine the draft status of all 19 year-old males. It was the first time such a procedure had been used since World War II. Priorities were established according to a person's birthday. Each of the 366 possible birthdates was written on a slip of paper and put into a small capsule. The capsules were then put into a large bowl, mixed, and drawn out one by one. By agreement, persons whose birthday corresponded to the first capsule drawn would have the highest draft priority; those whose birthday corresponded to the second capsule drawn, the second highest priority, and so on.

Theoretically, if the capsules had been adequately mixed the resulting sequence would have been random with respect to month. That is, a birthdate in September should have had the same chance of being selected early in the draft as a birthdate in, say, March. When the results were finally tallied, though, that didn't seem to be the case. There was an apparent tendency for days late in the year to be picked early in the draw.

Objective

To test whether the 1969 draft lottery was random.

Data

Table 1 shows the order in which the 366 birthdates were drawn. The first date, for example, was September 14 (001), the second, April 24 (002), and so on.

Analysis

We can think of the observed sequence of draft priorities as ranks, from 1 to 366. If the lottery is random, the average of these ranks for each of the months should be approximately equal. If the lottery is *not* random, we would expect to see certain months having a preponderance of high ranks, and other months, a preponderance of low ranks. This suggests that the lottery can be tested for randomness by using the Kruskal-Wallis statistic.

Question 6.7.1

Compute H for the data of Table 1. Test the hypothesis that the drawing was random. Use the $P = .05$ level of significance.

Table 1
1969 Draft Lottery, Highest Priority (001) to Lowest Priority (366)

Date	Jan	Feb	Mar	Apr	May	Jun	Jul	Aug	Sep	Oct	Nov	Dec
1	305	086	108	032	330	249	093	111	225	359	019	129
2	159	144	029	271	298	228	350	045	161	125	034	328
3	251	297	267	083	040	301	115	261	049	244	348	157
4	215	210	275	081	276	020	279	145	232	202	266	165
5	101	214	293	269	364	028	188	054	082	024	310	056
6	224	347	139	253	155	110	327	114	006	087	076	010
7	306	091	122	147	035	085	050	168	008	234	051	012
8	199	181	213	312	321	366	013	048	184	283	097	105
9	194	338	317	219	197	335	277	106	263	342	080	043
10	325	216	323	218	065	206	284	021	071	220	282	041
11	329	150	136	014	037	134	248	324	158	237	046	039
12	221	068	300	346	133	272	015	142	242	072	066	314
13	318	152	259	124	295	069	042	307	175	138	126	163
14	238	004	354	231	178	356	331	198	001	294	127	026
15	017	089	169	273	130	180	322	102	113	171	131	320
16	121	212	166	148	055	274	120	044	207	254	107	096
17	235	189	033	260	112	073	098	154	255	288	143	304
18	140	292	332	090	278	341	190	141	246	005	146	128
19	058	025	200	336	075	104	227	311	177	241	203	240
20	280	302	239	345	183	360	187	344	063	192	185	135
21	186	363	334	062	250	060	027	291	204	243	156	070
22	337	290	265	316	326	247	153	339	160	117	009	053
23	118	057	256	252	319	109	172	116	119	201	182	162
24	059	236	258	002	031	358	023	036	195	196	230	095
25	052	179	343	351	361	137	067	286	149	176	132	084
26	092	365	170	340	357	022	303	245	018	007	309	173
27	355	205	268	074	296	064	289	352	233	264	047	078
28	077	299	223	262	308	222	088	167	257	094	281	123
29	349	285	362	191	226	353	270	061	151	229	099	016
30	164		217	208	103	209	287	333	315	038	174	003
31	211		030		313		193	011		079		100
Totals:	6,236	5,886	7,000	6,110	6,447	5,872	5,628	5,377	4,719	5,656	4,462	3,768

Question 6.7.2

Compute the rank average for each of the months and plot those numbers on the graph below.

Question 6.7.3

Divide the 12 months into two groups, January through June and July through December. Do a Wilcoxon test on the rank sum for the two groups. Use the $P = .001$ level of significance.

Question 6.7.4

How would you interpret your answers to Questions 6.7.1, 6.7.2, and 6.7.3? What might have occurred that could account for the apparent non-randomness of the lottery?

REFERENCE

1. Selective Service System, Office of the Director, Washington, D.C.

A